STEVE GATENSBURY

ONCE, TO LEARN IT

A Lighthearted Account of a Fifty Year Adventure in the B.C. Lumber Industry

Illustrations by Dave Watland

Published by Sedge Publishing
1364 Beach Grove Road,
Delta, B.C., Canada V4L 1N6
Phone (604) 943-9740

Books are available from the publisher at $12.50 plus $1.00 postage and handling in Canada.

Illustrations by Dave Watland

Canadian Cataloguing in Publication Data

Gatensbury, Steve, 1923 -
Once, to learn it

ISBN 0-9694305-0-7

1. Gatensbury, Steve, 1923 - 2. Lumbermen - British Columbia - Biography. 3. Lumber trade - British Columbia - Anecdotes. I. Title.
HD9764.C22G37 1990 338.1'7498'092
C90-091104-2

Printed in Canada by Calli Printers Ltd.

TABLE OF CONTENTS

This book is dedicated to Ward Allen, a true friend, an amiable man, and an enduring credit to the lumber industry.

ACKNOWLEDGEMENTS

Thanks must go to my real word processor, Darlene Day, who helped to organize my work and who coped so calmly with my many afterthoughts and rewrites. And to Alan Dawe, my editor, who banished my superfluous asides and trite clichés, assuring me that with each pound of flesh he hacked away, he would not spill a single drop of blood.

I regret the exclusion of the names of so many of my friends in the industry. To mention some and not others would be unwise, and so I have named names only where they showed up in a story. It is sufficient to say that there were many who were the essence of sanity and consistency in sometimes circus-like circumstances. Leaving it at that allows everyone to say, "I think he's talking about me!"

FOREWORD

This book is a highly personal and quite incomplete record of a fifty-year romp through the lumber industry. It is not a book of facts and figures, and wasn't intended to be.

I covered a lot of ground and though I didn't play in the big leagues, I eventually did become an entrepreneur in the remanufacturing side of the business. I kept a low profile on the route I took, but it provided an interesting view just the same. It was a look from the underside.

Like most people, I lived a life of many incidents but few great moments. My formal education was interrupted early. I read too little over the years, mainly due to my preoccupation with work, and now I must admit that I have hardly a nodding acquaintance with the sciences, economics, or the details of the marvellous new developments of the world of high technology.

On the other hand, experience and maturity can be educators, and both are hard to avoid in ordinary day-to-day circumstances.

As an operator, I was always more concerned with the concept of where we were going and what the result should be than I was with the details of how we got there. I hired people to do details. After a while, through exposure, but without ever becoming a technical expert, I got to know how things should work and whether people and machines were working at their practical limits of efficiency. I realize that that sounds too easy, but it was a system that had its merits.

Too often when the experts that I watched let their projects get too intricate and sophisticated, they overlooked the obvious and had to rebuild parts of their major installations in order to get them to work effectively. They either lost their sense of proportion or they expected the raw material to accommodate their machine, instead of the other way around. I liked to start at the finished product and back into a scheme that would make exactly what we wanted.

It is true that what matures us are the well-remembered ordinary incidents of our lives, and it is the off-the-cuff, one-line observations that people make along the way that add spice and make life memorable. These incidents are, in the end, more educational than all the elegant theories on human behaviour put together.

Memories, of course, are subjective. This was clearly illustrated for me one day several years ago when my wife and I were going to Victoria on the ferry. After we had been sitting in one of the lounges for a while, Gerry asked discreetly, "Who is that fellow over there? He's been staring at you for the longest time."

I glanced at him and thought he looked vaguely familiar, so I went over and asked him who he was and where we might have met before.

Thirty years earlier, when we were both very young, we had sailed together in a ship that had the misfortune to be torpedoed and left stricken in mid-Atlantic. With this established, we fell into an easy conversation. Gerry, the innocent bystander, couldn't help noticing how different our recollections were and said, "You know, it almost sounds like you were on two different ships in different wars".

Our reminiscences did find some common ground when he recalled a brazen B & E job.

"Were you there," he asked, "when the door blew off the rum locker and a cask landed right beside the cups?"

I would like to have said that I lit the fuse, but I didn't, and besides, I think the lock fell victim to an axe.

My stories are as true as I can remember them but I won't be offended if you change times, distances, locations and players here and there to suit your own capricious memory.

Now, as the saying goes, a funny thing happened on the way to my retirement.

Chapter 1

Feasting With The Dinosaurs

I was shanghaied into the lumber business. To describe how that came about, I am going to have to tell you something about my father. My father had an interesting life but success just missed him.

My father and I were always at an awkward age. When he could have given me a lucid account of his times and travels, I was too young to appreciate them, let alone record them. When I was old enough, in the right frame of mind, and had a war and a marriage settled into their proper places, he was gone.

His name was Ernest Gatensbury and he was born in 1870. He came to America from Shropshire, England in 1888. He was eighteen and his only worldly knowledge was what he might have picked up living away from home in Devonshire where he apprenticed as a greenhouseman — not exactly challenging work. His father was a farm bailiff on a nobleman's country estate; that is, he collected the rent from tenants and managed the farm business. Thus father, though not from wealthy parents, was at least spared the horrors of the city life of those days.

In later years, I have been much impressed with this fact of father's tame beginnings, since the risks and scope of his adventures now seem to be so daring — at least compared to those taken by today's wandering kids for whom civilization is never that far away.

One of my cousins in England recalled the story of his departure. Apparently it was done in a very matter-of-fact manner. Probably to avoid objections, he announced his intentions on the eve of his leaving and his father said, "Well, I guess you'll want a ride to the station in the morning, then." I got the impression that he left with little more than the clothes on his back and what he could carry in his pockets, to try his luck in America."

This unsophisticated English country boy landed in New York City and I remember him telling of his trip across the continent by train. The transcontinental railroad had become a reality in 1869 but I guess parts of the trip were still through raw country. He told of the cars in his section of the train being left at a junction in the middle of the night in the middle of nowhere on the midwestern plains. They were, of course, picked up by another train which came along later in the morning but his first impression on waking was that they were stranded on a bald prairie that stretched right over the horizon.

His destination was San Francisco, and here he tried to get work as a gardener and greenhouseman on the city work crew. This was not easy as he was a foreigner with a funny accent, and knowing him later, he probably thought that eventually they would learn to talk like him. He was nicknamed "the lord," though this was surely for his English speech and manner, and not for any pretentions of nobility.

What he eventually did for a living there, I am not sure, but I presume that he was a gardener. I know he lived in boarding houses with other single men. I know he dated girls and took them to musical entertainments. One of these outings was to see Nellie Melba in concert, which must have been a rare treat for she was the toast of the musical world at the time. He used to sing bits of the hits of his day: "Daisy" and other Gay Nineties waltzes, Gilbert and Sullivan pieces and one that went "through all my days/I'll sing the praise/of brown October Ale."

After a couple of years, he left San Francisco and went to Pomona, California, which is on the fringe of Los Angeles. When he was reminiscing, places like Glendale, Pasadena, Anaheim and San Bernadino would drift in and out of his narrative. He lived around Los Angeles until 1895. While he was in Pomona, he worked on the construction and planting of Griffiths Park in Los Angeles. This is still one of the city's major parks today.

Later on, he had a bicycle shop in Pomona in partnership with a man named Louis Fox. He used to tell us about his triumphs as a bicycle racer, but since by this time he didn't look the part, we took this with a grain of salt. Even mother would scoff and say "tell us another," much to father's annoyance. But one day in the 1940's, cousin Fred in England unearthed a photo of father and sent it to me. There he was, young and handsome, astride his bike in a little cap with a visor, a white jersey with short sleeves that came down to the elbow and short pants that came discreetly to the knee. Despite all the coverup, however, he was a fit figure of a man with muscles in all the right places. On his chest were some medals on ribbons so I guess that at some level of competition he was a champ after all.

Father never did become an American citizen. One of the reasons he gave was the way he was approached by workers for both the Democratic and Republican parties urging him to join. Part of their enticement was the offer of a "shortcut" to citizenship. He felt that they were buying his vote and this offended his sense of propriety.

He said no to both parties and to being an American in the process.

Father was a very sensitive man. He read books, was well informed, and had been given a good basic education at the National Church School (Church of England). Why he didn't meet and marry a girl when he was a young man I have no idea. In later years, we used to tease him that somewhere there was another family that we didn't know about, but if there was, they never appeared. As far as I know, his first venture into marriage was when he was forty-nine years old.

Needless to say, I have pangs of regret from not knowing more about his life in California and even more over the period that followed, because in 1895 he got gold fever. He and his partner sold their bicycle business and went to the Klondike. This would have gotten them there three years ahead of the main rush of '98. Reports of people finding gold had been coming out of the North for several years before George Carmack hit the real stuff on Bonanza Creek, and there were a lot of people up there, looking in the wrong places.

I know that father went in over the Chilkoot pass from Skagway, Alaska and suffered all the hardships that went with the trip. His stories featured all the familiar names — Fortymile, Eldorado Creek, Bonanza Creek, Hunker Creek and the rest — because he was there for thirteen years. Many of the gold rushers didn't even get to dig a shovel full of gravel before they left and went home. Father did, and judging from his long stay, I assume that he made worthwhile earnings even though for the most part he worked on a share basis on claims that had been staked by others. He must have taken a less hysterical approach to gold digging than some of the others, since he came out twice for visits home to England before coming out for good in 1908.

The actual business of finding gold was not very glamorous. It consisted of sinking shafts through frozen muck and gravel to get down to the bedrock where the gold had settled during the grinding glacial action of the ice age. To get through the frozen ground, they had to build wood fires and thaw it out a bit at a time. The material that came out of the bottom of the shaft was brought to the surface and accumulated and became their "dump." In the summer, when the creeks thawed and the water was flowing, this material was put into sluice boxes that had running water diverted through them. The gravel washed off the top and the coarse gold and the heavy black sand, which also contained gold, sank to the bottom and lodged against crossbars known as riffles. This material was cleaned out from time to time and panned in the creek to recover the gold. In truth it was usually a whole lot of work and a little bit of gold unless you were one of the lucky ones who struck it rich and made a million dollars in a few months.

These men worked soaked to the skin in muck up to their waist, muck that was just off freezing, breathing wood smoke all the while. When their socks wore out, they "jumped" them — turned them over and wore out the other side. They lived a grim, Spartan life through endless hours of sunless winter weather, all in the hope that in the coming spring and summer they would get rich.

When I was in England during the Second World War, I was given some newspaper clippings that were excerpts from letters father had written to his parents from the Yukon. One was dated October 20, 1896, and the other September 16, 1897. Merely talking about them could not do them justice so I am reproducing them here in father's artless narrative flow, along with his erratic punctuation and the newspaper's comments:

LETTERS FROM A SHROPSHIRE MAN IN THE KLONDIKE

These are portions of two letters received by Mr. Gatensbury, of Moreton Wood, from his son, who has found his way to the Klondike. One of them was written in the autumn of 1896 and the other 12 months later, both having been nearly 6 months on the journey to England. They afford a graphic description of the exigencies of travel and the conditions of life in that region.

Fortymile Post,
Yukon River,
North West Territory, Canada
October 20, 1896.

Dear Father -

You will think I am a long time writing you, but as I have been living 60 miles away from anywhere with no writing paper or envelopes on hand, I could not write sooner. I have been living on Miller Creek all summer, and am going to winter there. I am down at the post getting some grub for the winter and have got work for all winter in a mine here at six dollars a day. I worked last summer for thirty days at a wage of ten dollars a day, which in English money is a little over £2 a day. But grub is very high here, flour being worth £2 a cwt. and everything else is in proportion. All the creeks are froze solid and have been since the first of the month. During the winter the thermometer gets to 70 degrees below zero. We have already had eight below. Fox and I came in together, but we separated. I am better off now than I have been since I was ill. I am out of debt, and have some money in my pocket, and enough provision to last me till next April — not bad for thirty days work. The wages are a fixed price all over the country — ten dollars a day in the summer and six in winter.

I have a nice warm log cabin to live in and plenty of wood handy for burning. If I make enough money this winter and next summer I may take a trip home in the fall and winter in England and come back for the following summer.

I hope that you are all well. I have enjoyed the best of health since I have been up here. It is the loveliest place in the world. In the summertime, the scenery is something wonderful. Ask Frank to get me all the "Tid Bits" he can get hold of and send them in care of the Alaska Commercial Co. They want to be at San Francisco by the latter end of August or 1st of May and then they would go by the company's steamboat as the overland mail does not carry newspapers. Reading matter is very scarce and the winters are very long lasting some nine months in the year.

They haul all the goods up to the creeks with dog teams and charge all the way from five to 30 cents a pound. It depends on how far they have to go. To Miller's Creek they charge 12 1/2 cents a pound. My grub bill for the winter is about 150 dollars hauling about 60 quite an item.

It is an awful long way to get here. We came from Los Angeles to San Francisco, then to Juneau by boat; from there we had to take the overland route, about 100 miles over a 3000 foot mountain and haul our supplies along. It was quite a pull to get our grub over the mountain. From there we built a boat and followed the Yukon River for 700 miles to Fortymile Post. At one place the river takes a chute into a canyon about 60 feet wide, the current running about 16 to 20 miles an hour. Below the canyon is what they call the White Horse Rapids. Twelve men lost their lives trying to run them. We let our boat down with a rope. We were very fortunate and made the trip without an accident. It will be a considerable time from now before you get this as it will have to be carried back the way we came in the dead of winter.

The Fortymile diggings are in British territory and right glad I am to be back on the old soil as the majority of Americans are insulting where anything English is concerned, but here they have to be careful. We have a fort and 25 mounted police.

There is lots of whiskey drunk here, as I suppose there is in any moving camp. They are considerably short of grub this winter. They will only allow 100 lbs. of flour to the man, which is a very small allowance for eight or nine months. You ought to see me make bread. I'm a dandy. I do all my own cooking, which mainly consists of black eyed beans and bacon without the beer with aleish taste.

Although we do have genuine Bass bottled ale and Guiness stout for sale right here in Fortymile and right good it is. It is 50 cents or 2s.2d. a glass (wine glass) rather an expensive luxury. I think I have written you enough. With love to dear mother and all the boys, I remain your loving son,

E.E. Gatensbury.

Second Letter

Miller's Creek,
Fortymile Post,
Yukon River,
North West Territory, Canada
September 16, 1897.

Dear Mother -

Your very welcome letter to hand. I was very glad to get it and hear that you and dear father are still well. I hope you will continue so. The winter has started here which is a little earlier than last year. It makes an awful long winter till next May.

I am down to the Post here, getting a winter's output of grub. I did very well so far, much better than I could have anywhere else.

A young Englishman and myself bought a claim and paid

750 dollars for it. It is turning out pretty well. We expect to have quite a nice stake by next spring. It was not so very cold last winter, 50 below zero that was all. We are likely to get a much colder one this winter. It goes down to 75 below sometimes. We live very well here considering. Lots of canned meat and vegetables, bacon and hams, condensed milk and other stuffs. We are just two on our claim our own bosses. When it is too cold we will just keep the cabin warm and stay inside. Some of our ground paid as high as 70 dollars a day apiece (£14 a day) We have got some pretty good ground to burn out this winter. We expect to take out any way £500 apiece. I have a very nice young partner. We get along first rate. He is about 23 years old, a rector's son from Banbury Cross. There has been some very rich strikes up here this summer and last winter. Men who had not a cent last year are worth 50 to 100,000 dollars now. It is supposed that one man alone will take a million dollars out of his claim this year. This is the first letter I have received from you since I left Los Angeles. Tell Frank...I intend to take a little trip somewhere with him when I come home, which won't be very long now. I hope this finds you all very well and happy. Give my love to all the boys, not forgetting yourself and father — from your loving son,

E.E. Gatensbury

Don't think I would have done better staying in England; far from it. I am doing very well and would not alter my course for anything.

E.E.G.

Father couldn't swim, but to start a visit to the outside world, presumably home to England, he made a trip on a raft down a river that was in flood. The raft hit a rock and overturned and father lost everything but his life and the clothes on his back. Included in his loss was a year's earnings in gold, so instead of coming out, he just went back in for another year.

In the winter of 1896, he and the men he worked with all placed bets on the fight between Bob Fitzsimmons and Jim Corbett. It was to take place in March of 1897 but they had to wait until the end of summer before they found out who had won.

Father had two close friends, Clem Blytheman, an English minister's son, and a Welsh seaman, William Fender, known as Coal Oil Billy. The origin of this unusual name is hazy, but it seems Billy had a scheme to buy up every five- gallon coal oil can he could get his hands on. He then cut the ends out, flattened out the remainder and used these sheets to shingle and weatherproof his cabin.

Father once brought Billy a good bottle of rum from Seattle to help him recall his warm days in the Indies. Billy said it was not bad, but he liked a drink that "burned all the way down."

Clem was a marksman and once split a shovel handle at a respectable distance to impress some native Indians with whom they were having a disagreement. I still have my father's Klondike rifle, a six-shot semi-automatic gas-reload Winchester 351 Cal. He told me that it had been a good provider.

Billy Fender gave Father's stories unsolicited corroboration when, sometime in the forties, he wrote a letter which arrived right out of the blue. This was the first time Father had heard from him since the 1890's. Billy was living in a seaman's home in Swansea, Wales, but his reminiscences of their lives and times tracked perfectly with my Father's.

Father told about the women who went up and down the creeks, staying with the miners, doing a little cooking and mending and whatever. "Of course I wouldn't want you to repeat that to your mother," he said with a veiled suggestion that went right over my innocent head at the time. Much later he put it plainer than that, but it didn't bother me, given the circumstances they lived in. He claimed not to be impressed by beautiful mountains because the beautiful mountains he saw for thirteen years blocked off his view of the sun that for a large part of the year was circling the horizon just below their peaks. Fortymile was above the 64th parallel and just 140 miles from the Arctic Circle.

At one point Father and his partner had several husky dogs and he told of how in a particularly bitter cold blizzard they thought they would show some pity and bring the dogs into the warm cabin. In minutes the dogs had had enough and asked to be let outside again where they then dug themselves down into the snow for shelter and

made themselves comfortable.

Father's best story concerned a steamer trip he made from Skagway to Seattle during which he shared a cabin with the notorious Soapy Smith, who was a conman, crooked gambler and gang leader and who, for a time, had the last word on everything that happened in Skagway. Smith was eventually shot by a member of a vigilante party in a shootout on a wharf on the Skagway waterfront. Father said he was rather pleasant company.

On one of his other trips to the outside, he made the 1700 mile trip down the Yukon River to St. Michael, Alaska, on the Bering Sea.

While he was home in England from the "goldfields" he spent his money freely, lavishing gifts on his mother and other relatives and on some of the young ladies of the village. Each time he went back into the Klondike, he had little more left than the price of his passage. How he squared this with his protestations that he was doing better than he could have done in England, I am not sure. His five brothers were all establishing themselves in businesses and all went on to become quite comfortably successful. Perhaps they, in turn, envied him his free-wheeling life.

In 1908, Father left the Klondike and went to Ontario to the area around Kenora, which was then called Rat Portage. He sank his capital into a hardrock gold mine venture that proved to be little more than a hole in the ground. The gold bearing vein they were working stopped abruptly at a fault in the rock and disappeared. He was unable to raise enough money to continue the development and so he had to go out of business. He had progressed to the point where he had installed an ore mill which he had financed with bank assistance. When the gold stopped flowing, the bank called its loan and blighted father's opinion of bankers forever.

With what he salvaged from this venture, he arrived in B.C. in 1910. For the sum of $1,250 he bought five acres of land fronting on Como Lake in Coquitlam where he built the greenhouses that were the forerunners of all those that were to follow on that property until 1988, and he got his name established on our street. The minutes of a Coquitlam Council meeting of March 12, 1910 recorded the following: "E.E. Gatensbury, et al. asked for some outlet from their land at Lake Como. On motion of Councillors Fox and Tyner, this was left to the Board of Works with power to act."

Once established, the greenhouses became a well-known business in the area. Two of his hired men were Lucien Canuel and Jimmy Hamilton. They were two of many as it was a business with peaks of activity. When he needed help, dad rounded up the neighbourhood kids and some from down in Maillardville. Hamilton told of working for $9 per week, which, compared to Father's $10.00 per day in the Klondike, sounds like slim pickings.

These must have been free-wheeling days for him. Twice a week he would rise early in the morning and make the long trek from "the lake" to Vancouver along the Barnet Highway to the warehouses of the produce wholesalers on Water Street. Other days he went to the city market in New Westminster. These trips would see his wagon piled high with crates of lettuce, cucumbers and tomatoes. He was determined that these were the things to grow even though his contemporaries, Dwyer, Whiting and Tidy, had already changed to the more year-round business of flowers and bedding plants, abandoning the vegetable market to the burgeoning Chinese truck gardens.

When the business of the day was done, the rest of his time in the city became a social ritual. He was with familiar friends and associates whom he met on a regular basis in the city. Father was known to take a drink, and after an extended visit at a favourite watering hole he would put himself in the hands of Tommy, his horse, who

knew the way home. I remember him saying that the distillery on Braid Street even extended its hospitality to thirsty travellers in those days in much the same way that the breweries did for the men in blue and other privileged characters for so many years.

Living at the summit, so to speak, Father had the option of going down the hill to the Maillardville side or to the Port Moody side for this social life, and he had friends in both areas. One of my earliest memories is of visits made to a family named Aston who lived in a houseboat on the inlet at Port Moody. From the plank walkway around the house, we watched jellyfish in the water with great fascination.

Like so many others, Father was in the right place at the right time, but had no way of knowing it. In its wilderness state, with bears almost matching humans in numbers, the area could not have been prepossessing. But at $250 an acre, anyone with a few thousand dollars could have bought the whole hilltop; it was practically vacant. (Just imagine all those elegant homes that fan out there today in all directions within a mile of his old property, then count the millions.) They gave the street his name because he made a commotion to have it built into his place. This was by no means an exclusive honour. Many, if not most, of the original street names in the area came about in a similar manner; that is, the inference was that it was Gatensbury's or Porter's or Foster's road.

He told many stories about the tireless Dr. Scott, the deadly flu epidemic of 1919, ice on Como Lake and the several times he was called upon to use his ladder or some other makeshift gear to rescue someone who had tempted fate and fallen through.

Somewhere around this time the stories become more familiar to me.

My mother came to the West Coast in 1918 to visit her sister who lived on Blue Mountain Road in Coquitlam. Mother was from Nova Scotia, but she had spent the previous ten years in Boston.

Father was one of many bachelors who lived in that early community.

Dora Holroyd, Annie Walls and my mother were the only young unattached women in the area.

Father pressed his attentions on Mother and by the more particular rules of that day, once they were seen to be keeping company, the next move was almost obligatory. Certainly the situation did not lend itself to her playing the field, reputations being as precious as they were.

My mother and father were married in 1919 by the Rev. Frank Plaskett in St. Mary's Anglican Church in Sapperton. He was forty-nine and she was twenty-one years his junior. Father was comfortable with his cat and dog in a rather nice little house, but the day Mother arrived, the cat left.

Mother brought her rock-hard Nova Scotian Baptist principles with her and this brought about a radical change in Father's style. He always made a little berry wine, but I don't think I ever saw a bottle of hard liquor in our house when I was a kid.

They met people at this time who became lifelong family friends. They attended the "fair" at New Westminster and to Mother's consternation, if not horror, they went to those fierce and famous field lacrosse games between New Westminster and Vancouver. This had been one of my Dad's regular outings before he was married, and the sight of E.E.G., the rabid sports fan, was more than she had bargained for. It was expected of a faithful Salmonbellies fan that he uphold the team's and the city's honour, even if it meant the occasional bout of fisticuffs to set the other fellow straight.

To return briefly to the Coquitlam bachelors who were mentioned earlier, a friend was able to "one up" me on the subject of September-May alliances. She had gone through her life thinking that at the time of her parents' marriage, her mother had been twenty and her father forty-five. Her father had made it known that he wasn't keen on having any children and my friend said, "Small wonder. When my mother died and I rummaged through her things, I discovered that my father was actually fifty-nine when he was married and sixty-five when I was born." My sister added a further dimension, recalling that the bride in question was, in truth, sixteen not twenty!

In the summer of 1920, Father built a number of new greenhouses and had them finished except for the heating plant, which was months late in delivery thanks to plumbing and heating contractor Grumpy Spring, his erstwhile lacrosse hero. He had built the new greenhouses as large, single compartments, each covered by a series of pitched glass roofs which were joined together like saw teeth. This was intended to make the heating arrangement more efficient. While the greenhouses sat there waiting, there was a freakish snowfall of nearly three feet, which was followed by pouring rain. The resulting mixture could not run away fast enough. Even though Father and his hired men ran through the buildings breaking the glass to let the slush fall through, they could not do it fast enough. The houses collapsed. The business collapsed with them and a lot of the spunk went out of Father as well.

Since the project was financed by loans, the property went to the mortgage holder. Some of Father's other creditors were paid off over a long period of time. Duns and threats from some of the others would occasionally appear on the impressive letterheads of legal firms but for these creditors it was over too.

One of Father's early backers was the Roe and Abernethy sawmill (Canadian Pacific Lumber) in Port Moody. This company became Abernethy and Lougheed, then Thurston Flavelle, then Flavelle Cedar, then Canadian Collieries, eventually becoming Weldwood of Canada, which now operates on the original site.

The trigger on this whole disaster was only ninety-seven hundred dollars (in 1921 dollars), but it was enough. My sister recalls that she made the last payments on two of these accounts in the early 1940's. One was reducing at two dollars a month, and the other at seventeen dollars a month and change.

Father needed the firm hand of a saving woman and Mother always insisted that if she had arrived on the scene a few years earlier, things would have been different. She was probably right, because her management of the family's financial affairs from then on was strict and effective.

The remainder of Father's life was no great adventure. He now had family responsibilities, and he worked hard for wages in a sawmill until he was nearly seventy. Only once or twice do I remember him sighing that he should have remained a bachelor.

The "geranium of hope in the window" was always his garden. After a brief stay in the company townsite at Fraser Mills (where we were living when I was born), we made our last move. We had acquired an acre of land in Queensborough on Lulu Island (it was actually four lots bought at $50 a lot), and this became Father's pride and joy. He cultivated every inch of it, and his efforts earned him the respect and praise of everyone who passed by. He gave his original trade the touch of an expert.

The move to Queensborough was somewhat traumatic for Mother — something she was learning to live with. My dad had unexplainably bought the property without telling her, and they moved into a one-room shack which served as home until they built their house.

It was not the thought of starting over that dismayed Mother: it was because she was born and raised in a "hollow" and had a real aversion to fog and to swamp life. (She soon overcame this and became well-known as the first president of the Queensboro P.T.A. and also for accommodating a "well baby" clinic at our house for several years.)

This shack was down to earth basic and one of its features was my bed. I slept in something known as the "rag box," a narrow, rough, wooden box filled with blankets, pieces of cloth and old clothes. I was so confined in this that I am told I complained loudly that there was no place for my elbows.

The new house, or "half a house," was built by Mr. Sawyer, an old Como Lake friend whose carpentry skills seemed to be rudimentary at best. The first section was certainly not an adaptable module. The family story is that nearly the whole thing had to be rebuilt a few years later when the time came to complete the house and make it into the comfortable home of my youth. At any rate, we moved into the first phase, shiplap and tarpaper on the outside and exposed studs on the inside, on a cold, rainy Easter Sunday in 1924, and settled in for a long stay.

When Mother and Father first arrived, it didn't take long for the disparity in their ages to catch the neighbours' attention. The story went around that the new neighbours were an older man, his daughter, and her two children.

One of Father's few forays out of the range of mother's watchful eye was on the night that R.B. Bennett and his Conservatives were defeated in 1935. We had no radio or telephone, so Dad went into New Westminster to hear the news. He hit a couple of old familiar haunts and came home more than firm in his conviction that just when his man had got this depression thing straightened out, someone else was going to take the credit — an opinion shared by few others at the time.

Perhaps his only legacy from his gold rush days was rheumatism in his knees, about which he complained year after year. This caused him to walk with a rolling gait that made him recognizable from a distance to anyone who knew him.

He was a tough old guy and he never lost a day's work, even through the depression years. He worked at Fraser Mills, and sometimes when the company would get a big offshore order and the ship's arrival date would become their deadline, the mill would work from eight in the morning until nine at night, night after night, until the order was out. The people he worked with remember him simply as "a nice old gentleman."

In spite of the terrible times, we ate well and were comfortable on our little farm. In about 1936, I remember Mom and Dad having a playful ceremony on the front lawn when they burned the mortgage on the house and property.

One of the few interruptions to Father's endless days of work was the strike which occurred in 1932. This became a bitter affair that separated the old style loyal employees, who were thankful for their jobs, from a skeptical faction who thought that, because of a series of wage cuts, the company was being mean and heartless. Father fell into the first category, and when a delegation called at our home for a donation of food to help the hungry strikers, father brusquely sent them away, saying that if they wanted to eat they knew what to do. This gave him a reputation that, for the rest of his working days, put him on the wrong side of the young Turks of the growing union movement.

One night in May of 1939, father came home from work and announced that he just couldn't do it anymore. While he had operated a machine in the mill for many years, he was only doing a menial job at the end. Well, the fix was in. My future was planned. The next day I left school and went off to work at the same job he had been doing. He was 69, and was making sixty-five cents an hour. I had just turned sixteen, and got thirty-five because boys were paid less.

— * * * —

i) Greenhouses at Como Lake, 1918—1920.
Wilderness in the background.

ii) Mother in the lettuce beds at Como Lake, 1920.

iii) Father, home from the Klondike for visit, about 1900, with gold nugget stick pin.

Chapter 2

Out of the Nest Without my Feathers

Looking back to that time, I am satisfied that I was not given a choice because there was no choice to give. As soon as father was seventy, he got the old age pension of $20 per month, but even with that there was a means test. I'm sure that the government looked at our situation carefully to see if there was really a need. I have a vague recollection that somehow I was considered a means in their eyes.

I had not been at work very long when I gathered up my courage and went to see Maurice Ryan, the superintendent. I pointed out that I was doing the work my father had been doing the week before for roughly half of what he had been paid, while we still had the same household to maintain. He said he understood my problem but that if he gave it to me, he would have to give it to every boy in the plant. That was a pat answer that would not do today. But that was 1939.

— * * * —

At the time I started there, the Canadian Western Lumber Company operated Fraser Mills. Its trademark was a red Ⓕ and it was known all over the world. Fraser Mills was a huge sprawling plant with a sawmill, planer mill, dry kilns, veneer and plywood plant, sash and door factory, shingle mill, deepsea loading docks, and acres of yard. At its peak, there were fifteen hundred employees, and it cut a million board feet a day in the sawmill. The production of the rest of the complex — the shingle mill and the veneer and plywood plant — was in addition to this.

All this activity was timed and commanded by a plant steam whistle that blew in a grand throaty baritone and ended with a marvellous groan. There was no mistaking it.

At one time the B.C. Electric interurban train went down into the middle of the plant, but that was discontinued when I was just a kid, and by the time I went to work there, even the entry of private autos was prohibited. This meant that after a hard day's work, everyone had to trudge a quarter of a mile up a long planked roadway to the mill gate where some people got into their cars and others onto streetcars. By far the largest number continued to walk three and four abreast in army-like ranks, up King Edward Avenue into Maillardville, some of them walking a couple of miles to get home.

In 1909 and 1910, a total of 210 French Canadian workers and their familites had arrived in the area from Quebec. The men had previously arranged to get jobs at the mill.

Even at that early date, there were 400 people employed in the mill. It was famed then as the "largest mill in the British Empire."

Tony Pare made a matter-of-fact comment when recalling the story of those early days. "The 1910 contingent," he said, "included thirty-one Pares."

Within a year of the arrival of the French Canadians, a church had been built and the community was named after Father Maillard who was the first parish priest.

I was always impressed by the evidence of a reverse assimilation that gave us French Canadians in the mill crew, whose names were Finnigan, Ramsay, Thomas, McLure, and other Welsh, Scottish, Irish and English names. On a random tour through the plant, you would hear many conversations carried on in French.

Fraser Mills must have been the smallest municipality in B.C., but it truly was one. It comprised 390 acres, although much of this was wasteland — bog covered with hard hack — right up until the late 1980's. The company was its only taxpayer — everyone else rented from the company — but the reeve and the councillors were each given temporary theoretical possession of a lot which qualified them as property holders for the duration of their terms of office.

The municipality consisted of the mill site, a company store, a first-aid station that was like a small hospital, a "club house" which was a single men's boarding house, three ethnic communities, and about seventy-two company houses that stretched up King Edward Avenue to Brunette Street with three side-streets at the bottom of the hill. The millsite and the land at the bottom of the hill were bog — a quaking, quivering bowl of jelly — and every vibration was transmitted in all directions from its source. I can remember sitting in any number of homes half a mile away from the sawmill and hearing the dishes rattling in perfect time with the up-and-down strokes of

the big old Wickes gang saw in the mill. As well, when the sawyer would turn a particularly large log, the thump, when it fell onto its flatted side, would feel like a minor earthquake in the townsite.

Though there were many employees who came down from Maillardville and from New Westminster and beyond on the streetcars, the local residents kept their identity as a distinct community. Everybody knew everybody else in the townsite and there was a definite pecking order based on the quality and positioning of your house. The farther up the hill you were, the farther up the ladder you were.

There was a fire brigade and a fire alarm system that consisted of a coded series of blasts on the mill whistle. Most of the firemen lived in the townsite so they would be quick to respond.

At one time the company had an entry in the Vancouver senior baseball league, though the players were mostly American imports from the University of Washington where Maurice Ryan, the superintendent, had gone to school. Later, they had a team in the Dewdney League at a time when community sports could still rivet people's attention. This team reflected the true spirit of the neighbourhood, being about half French Canadian and half mixed others.

The company retained two doctors, Dr. Cannon and Dr. Sinclair. They had regular hours for seeing the employees and their families in the hospital on the company property.

Dr. Cannon was a gruff, unemotional man who had been an army doctor in the First World War. Not many will remember him for his sympathetic bedside manner, but I guess from the company's point of view, he was a one-man judge and jury on the legitimacy of aches and pains. Still, he could be ambivalent. He once told me that if you wanted to establish a sore back or a hernia as a work-related injury, you should lie down and make them pack you away.

— * * * —

Fraser Mills was a major cargo shipper and the company docks would often service two and three deep sea ships at the same time.

In the thirties and forties, most of the freighters were still around ten thousand tons displacement and would hold at most, eleven million feet of laboriously hand-stowed lumber. By comparison, the bulk loading ships of today will go up to fifty thousand tons displacement, and carry thirty-six million feet of lumber that is conveniently stowed by the ship's own bridge crane that picks the lumber off trucks which have driven through the stern right into the ship's hold.

In the heyday of the general cargo freighters, there were thousands of ships of all registrations, crew-nationalities, hull designs and propulsions. A few still burned coal, most were oil burning steamships, and a few of the classy refrigerator ships were diesel powered.

In the thirties, life on these freighters was tough. The shipping companies were struggling and they passed part of the struggle on to the crews. I have seen many enamel-ware mess trays with mutton stew in one end and a glop of rice pudding at the other, and the whole thing running together in the middle. Some of the Greek ships kept live sheep in pens on the upper deck to be slaughtered as needed.

There was even a story that Lord Runciman, who was a ship owner and, at the time, president of the British Board of Trade, could not get his ships refitted in England and had to take them to Holland. They were in such a rusted out state that no one wanted to touch them. A few rusty ones hit our ports and they added a new definition to the term "tramp steamer."

Being small, these ships were able to come up the river to New Westminster and then on to Fraser Mills where they loaded their cargos off the company docks. The seamen from these ships were a familiar sight as they walked through the mill property up the long walk to the streetcars to go into the city. Sometimes, on fine days, they played sandlot soccer on an area of filled land at the west edge of the plant.

When I was still in my teens, the chief engineer of a British freighter let me sail up the river with them from New Westminster to Fraser Mills. It was a big deal at the time because depression kids had to find their adventures wherever they could.

— * * * —

After I had grubbed along at my dad's old job for awhile, trying to find a way out of it seemed the natural thing to do. I had been recruited by Scotty Henderson, the kiln foreman, to do his moisture content calculations. They were beyond him but really required only grade five arithmetic.

The test consisted of drawing a piece of lumber out of a kiln pile, cutting off a very short section, weighing it, drying it bone dry in an extremely hot oven, and then reweighing it. The difference in weight was the water that had been cooked out of it in the oven and a calculation determined the moisture content of the material still in the kiln.

Scotty, who was an old uneducated fellow, had a different method which he preferred. He would rattle a stick down the side of the stripped load as you would along the pickets of a fence. To his experienced ear, the tone of the rattle told him how stiff and hard the lumber was and therefore, how dry it was getting to be. He would also smell it, knock it with his knuckles, and rub it with his hands.

All of this was later replaced with an instrument that measured the moisture content by electrical resistance. Two needle-like probes were driven into the wood and the degree to which a current could flow between them told how much moisture remained in the wood. It was as simple as that.

Scotty is now gone so my story is safe to tell. He built a fireplace at his Blue Mountain Road home. It was probably one of the most economical fireplace jobs in history. Scotty took home two fire bricks from the boiler house repair stock in his lunch box every night until he had enough to do the job.

— * * * —

Somehow, being old Scotty's helper didn't strike me as the start of a career and I soon found a way to make a move. Because of the fire hazard, smoking has never been allowed in the sawmill work areas, and these were still fairly unenlightened days when a fellow didn't get a legitimate break during working hours.

This led to a situation where people covered up for each other while they sneaked away for a puff. All this leads up to the fact that, though I never did smoke, when it was my turn to sneak, I sneaked up to a grading station where I made friends with a grader. While he was having his smoke, I did his job and began to apply the meager knowledge I was picking up by attending the industry-sponsored grading classes. The grader was Mac Ewart, who was later to become the resident manager of the mill. Mac left the company for awhile to become a foreman at the Macdonald's B.C. Manufacturing plant, and he tipped me off when he landed the job. I immediately went to Maurice Ryan to ask for Mac's job and he exploded, "Jeez, you know more about what's going on than I do. Mac hasn't told me anything!" I got the job. Seniority had nothing to do with it. I was Johnny at the rat hole and I knew how to do the work.

Another teenaged grader in that area of the plant at that time was Asa Johal, who now owns and operates one of the largest independent sawmills on the coast today, Terminal Forest Products.

— * * * —

There was not much in the way of amenities around the old mills. There was no lunchroom — you just sat down with your lunch box anywhere that you could get out of the wind. On nice days, we would go down to the

docks in the fresh air and share our lunch with the gulls.

If there were no lunchrooms, you can probably imagine what the toilet facilities were like. A cavernous room was built out over the river. Each side of the room had a row of triangular seats that looked like saw teeth with the points aimed at the centre of the room. There were no cubicles and when the wind was blowing (and it was always blowing under this setup), your backside became decidedly chilled.

— * * * —

Later, I started to spend all of my unofficial "smoke" periods in the sawmill so that I could see the logs being sawed.

Here is a diagram from one of the early grading class books. It depicts the ideal log — probably Douglas fir, four or five feet in diameter, round and sound to the heart. If you want to think of it as a meat cutting chart, the steaks are all on the outside and the pot roasts and stew beef run down the middle. It is saddening to think that, from the next generation on, no one will ever see a log like this in a sawmill again. It takes five or six hundred years to grow one, and I doubt if anyone is prepared to wait that long.

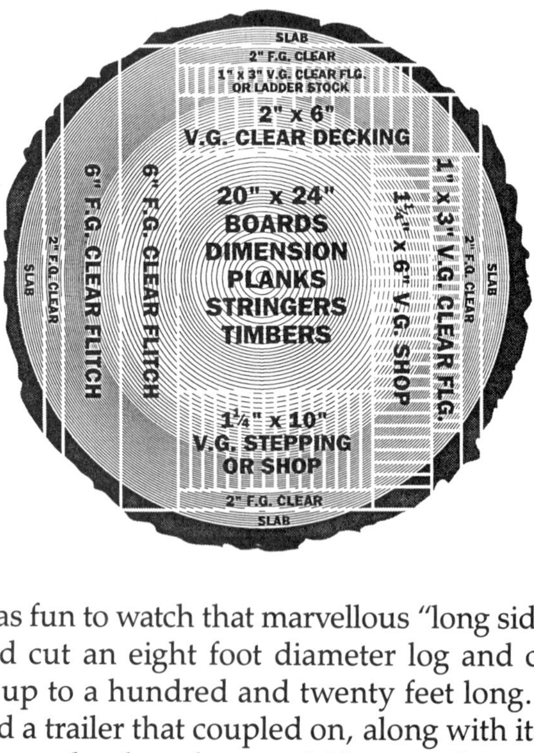

It was fun to watch that marvellous "long side" where we could cut an eight foot diameter log and could cut timbers up to a hundred and twenty feet long. The carriage had a trailer that coupled on, along with its air connections, so that long logs could be cut all in one piece. In case you're interested, some of these long pieces were used as dredge spuds. In this use, one end was pointed and metal clad, and several of them were stood on end in guides and driven down into the soft mud where the dredge was working to anchor it against the current of the river, and to steady it against the pressures of the dredging operation. Of course in earlier times long logs were also cut for masts and spars.

Other than this, we always had a sneaking suspicion that timber customers just bought long lengths to give them more options when they cut them back to short lengths in their own plants.

The log haul could not reverse and occasionally the boommen would get a log on the chain that was too big to go through the opening in the end of the mill. All they could do was cut the hole bigger.

On one occasion, they cut the hole bigger but the log was still too big to load on the carriage. The only thing they could do then was to bore holes and blow the log with stumping powder. This was a common method of splitting logs, but not as a rule on the log deck, right in the mill.

— * * * —

I guess we are all kids a lot longer than we think. We knew early the conflict between duty and human nature. One of the few things that could shut a mill down was a break in the chain that brought the logs up into the mill from the river. When this chain broke, it didn't just break; it broke and fell into the river at the bottom of the jack ladder. This meant much fishing to recover it and it was hard to handle when they did locate it. The metal in the links was more than two inches in diameter so it was very heavy.

At any rate, we would see the mill completely clean out with no new production coming behind it. Then the word would come through the verbal telegraph that the log haul chain had broken and we would all whoop and holler, even the old guys, because we knew we wouldn't have to work for the rest of the shift.

— * * * —

One of the problems of winter operation occurred when the weather turned cold and the log pond and the next boom of logs to be used froze together in a solid mass. The boom men would break a hole through the ice between the logs and lower a charge of blasting powder into the water. The secret was to have a good depth of water under the logs so that when the blast went off there would be an upheaval of a large area of water

and the freeing up of a lot of logs. I have tried it where the booming ground was shallow. All that this accomplished was the total destruction of three or four logs and the opening of a very small hole in the ice. (I suspect that these days the fisheries people would clap you in irons for even thinking of it.)

— * * * —

Running parallel with all this stuff was an introduction to life — working with mature men. I guess I was fortunate in that respect. Even though the work we were doing was dull, these particular men talked intelligently. It was probably to my benefit that I missed out on working with kids of my own age who might have had a wilder lifestyle than mine. When I look back, though, I can see that there were still lingering traces of the depression, and most of the jobs were being held by family men.

My job during my first year was to pile and strip lumber for the dry kilns, and since these were units of uniform size, it was very easy for the foreman to measure our production per hour or per day. There were Chinese workers in the crew and they paired off with each other. It was a strange thing how they seemed to take as much time to talk, throw orange peels and make rude gestures as we did, but no matter how we tried, we could never match their production.

Their secret was that they never really stopped. One hand was always out for the next piece, so they produced more, and what is more, it was more for less. I have just read Paul Yee's book, Saltwater City, which relates the story of the Chinese immigrants in B.C. and the way they were discriminated against both in pay rates and in blocked access to anything but the most menial jobs.

In my innocence, it had never occurred to me that the Asian races would be paid less than we were. I had to ask an old friend if this was true during our times there and he assured me that it certainly was. Still, some of them were able to jump up a notch and got better jobs. I don't remember any outward animosity toward them. It was true that they couldn't read or write English, but through good memories and mature judgement, they did grading jobs and operated machines as well as the next man.

The Chinese workers were mostly bachelors whose services were procured for the mill by labour contractors, Barney Kong and Mrs. Wong. These fellows would finish their last shift before the weekend, and in an hour or two could be seen in suit and tie on the streetcars headed for Vancouver's Chinatown where, by their own account, they spent the weekend eating, drinking, gambling and spending a little time with the girls.

— * * * —

There is no truth in the notion that all the old time bosses were hard boiled.

I broke a rule very early when I got involved in a scuffle on the job. The foreman had asked me to set up to pile some of the heaviest material in the range of things that we did. My partner, a mature man and partner of short duration, was busy setting up for the lightest material. When I told him of the foreman's instructions, in the true spirit of shoot-the-messenger, he made some unparliamentary comments about my relationship with the boss. For this I punched him in the mouth, and in return he hit me on the head with a kiln stick which hap-

pened to be within reach.

I was called to the office where Maurice sternly told me not to take the law into my own hands — to come and see him if I had trouble. He was probably saying under his breath, "Jeez, how did I ever get saddled with this bunch of idiots." Fortunately he didn't say "You're fired," or "Take the week off to think about what a dumb thing you did." The trip to the office was punishment enough; I was sure I had blown my job.

There was a young fellow in the kiln crew and for a time I was paired with him. Roy Mosure had an excellent ear for music, was interested in what was going on in the world, and was desperate because he *knew* his job was dull. It helped when you didn't know or didn't care. He welcomed me as a partner because our spirited conversation enabled us to blot out the truth that we were lousy substitutes for a machine.

Just a few years later, I was in a movie theatre in

Halifax and saw Corporal Roy Mosure waving to the people in the theatre from a newsreel taken somewhere in Europe. Sometime after that we heard that he was dead. He had been shot in the dark by one of his own men while they were on patrol.

— * * * —

One of the best remembered old timers at Fraser Mills was a mill foreman named Snap Stewart. These were the days before companies had become title happy. There was a general manager and a superintendent, with a mill foreman on each shift. Any paperwork there was got done by people whose titles were no grander than "tallyman" or "clerk."

The organization charts were very simple. The mill foremen under these circumstances ran the mill in every sense of the word. They even did their own hiring, which led to some incidences of kickbacks and other irregularities, but the foreman was boss and everyone knew it.

Snap Stewart was an American who had come up from Port Blakely and Bellingham in Washington. Besides upholding the sawmill traditions of smoking, drinking and chewing, Snap had a personal quirk that set him apart. He kept a journal. This document is priceless because it gives a firsthand account of the lives and times of Fraser Mills and its employees from 1905 until 1945.

I once had the opportunity to read this book. A visit with the book is a better description. For someone who had been there, reading it was a kick, because many of the names mentioned were people who were still around, some of whom I knew personally. The book was not a production record. The only details of that type were accounts of exceptional production days, exceptional sizes of orders or specifications, or unusual cargos. The earliest entries reported on sailing ships that had been towed up the river to load.

All of the shipments went by water until June 10th, 1912, when the journal reported the first rail carload shipment to leave the plant.

These were the days when wage rates followed the fortunes of the company up and down, and, I am sad to say, Snap's journal did not report too many "ups". The downs he reported:

Nov.11, 1920	10% cut
Dec.31, 1920	20% cut
Sept. 1, 1921	5 cents across the board cut on "common" labour
Dec. 3, 1926	10% cut
Mar. 3, 1927	A cut that Snap did not specify

In 1932 there was a 5-month strike that became a violent affair, but the employees still took a 5 per cent cut when they returned to work. Shortly after that, things did begin to improve and an occasional increase was given.

But there was also an entry on March 3, 1927: "For the first time in the company's history there are no orders for the mill to cut."

Most of the entries in the book were gossip — the day- to-day details of births, deaths, weddings, new cars, personal peccadillos, and the weather. My mother and father were living in the townsite when I was born in 1923 and my arrival was duly noted.

Snap had a pal, Joe Abbott, a native Indian —

Qualicum Joe. They were Saturday night drinking buddies and, in the journal, Snap quaintly lapsed into Stengel-ese to describe how Joe and "that other fellow" drove their car off the edge of the Penitentiary hill in New Westminster to tumble a hundred feet or so to the train tracks below. They both walked away unharmed.

Joe Abbott was a sawyer in the days when all size and setting information was communicated by hand signals instead of the remote controlled electronic devices or computers used today. The setter sat in a bucket seat, riding the carriage up and down the track, and, through mechanical controls, set the log out the right amount for the next piece to be cut off when the log passed the saw. The sawyer read the quality of the log, its size and its potential for the orders he had to produce. Joe could break down a big log, taking off various sizes, many of which were fractional. To the size of each piece he had to add for the saw kerf (the wood that was lost in sawdust with each cut), but when he got down to the last piece, it would always be the size he expected it to be. He wouldn't even have to take off a shim to make it exact.

Joe could not read or write, but he could add and subtract inches and fractions of inches in his head as the piece remaining on the carriage got smaller and smaller.

Joe was a notorious tightwad and grouch. He would never contribute to any collection for a fellow worker's wedding (let him pay for his own screwing), even though he was paid $1.25 an hour as a sawyer when the basic wage was twenty-five cents. When Joe was an old man in a nursing home in Marpole, a friend took him cigarettes, but he complained that they weren't his brand.

— * * * —

Christmas Eve at Fraser Mills was a pretty relaxed day. There is no way anything like it could happen today. There was a major refreshment centre set up in the blacksmith's shop and H.J. Mackin, the president, came down and had the first drink. Most people in the plant who drank brought a bottle to share a drink with a friend, and after a while the scene got a little wobbly.

Steve Jarrett was a teamster who seldom made it past the middle of the morning on Christmas Eve. His usual job was to go back and forth from the dry sheds to the planer mill with horse cart loads of dry lumber which the planers ran into flooring, mouldings and finished materials. By about 10 AM Steve was nowhere to be seen, but his horse would finish the shift on his own. The men

at each end of the route would hook and unhook full or empty trucks as required, throw the reins over the horse's back, slap his rear and send him on his way. At the other end, the same thing would be repeated. I have to assume that the foremen and superintendent knew what was going on, but that was a day when everyone looked the other way.

Life is full of unexplained examples of tortured talents. We had one, a tallyman whose name was Dick Haines.

This gentleman was an accomplished musician who had played concerts in Royal Albert Hall in London. He was also a remittance man, sent to the colonies on a retainer to save the folks back home from embarrassment.

Word got around about his ability and he was approached by a major church with an offer to become their organist. He said, "I would like very much to be your organist, but on Sunday mornings I am often drunk and I don't think your parishioners would approve."

— * * * —

It is hard to believe that, when I started to work, there were still quite a few horses working in the mills. Flavelle Cedar kept them longer than Fraser Mills, but we had them and we were familiar with their quirky personalities. That old familiar saw, "Hold 'er Zeke; she's headin' for the barn," was based on real experience. The teamsters would head to the barn at noon and at the end of the shift, riding an abbreviated chariot that consisted of the shafts, two wheels with the axle between them and a skimpy little platform that held it all together. When he was not hooked up to a load, the teamster stood on this platform, holding the reins.

Horses, having generous quantities of horse sense, knew when the last load was finished, and when the teamster headed for the barn, life and limb were in serious danger. The horse would take off at a fast trot with the teamster reining him in and this in itself was a skill-testing exercise, but occasionally the horse would break into a full gallop and then the teamster played Ben Hur, and had to be very good to stay on. The horses knew what

the whistle meant and at the suggestion of working a little overtime, they told the teamster where he could put it.

— * * * —

The planer mill had four moulders and four planers, a rip saw and a resaw, and that's a lot. In a separate building there was a yard sizer (ancient terminology — it was a fast Stetson Ross dimension planer) and, in addition, there was a timber sizer at the skids that could plane up to 16″x16″. I have a reason for going on about this, which I will get to.

I was always fascinated by the dry sheds. In those days, the mill loaded out mixed cars containing everything from timbers and framing lumber down to the smallest screen moulds, coves, quarter-rounds, casings, baseboards, stepping, handrails, and even wood gutter, not to mention the sash and doors and plywood. The tallymen had to figure out the cubic feet of space required by each item and plan so that he got all his order into the boxcar. He had a crew with a pushcart that went up and down the alleys in the sheds, picking up just the right amount of each item, which they then brought to the car door for loading.

Here is the point I was coming to: nobody at the time thought much about remanufacturing, secondary manufacturing, or "adding value." It was taken for granted that these were part of normal operations. But there came a time when the companies tore down their sheds and elaborate planing mills, and dismantled their dry kilns in a swing towards simple, high-volume cutting programs.

One could be forgiven today if he thought he heard recurring strains of the song "Everything Old is New Again," as the renewed interest in "value-added" products intensifies.

— * * * —

Throughout my working life I have always had the notion that the line that separates work life and social life is a fine one. For most people, arriving at work provides a certain satisfaction. Many would deny this, but when it is taken away they miss it acutely. It derives from familiar faces, familiar circumstances, friendships, and continuing conversations. People at work are alert, their minds are focussed, and their memory is better there than in any other place because everything around them is a reminder. For the most part they are dealing with situations they can handle, and they are comfortable doing what they know how to do best. I can say that this was true for my fifty working years and I went out of my way to promote this notion later on when I ran my own business.

Here is an example of this family feeling at the Fraser Mills of those many years ago. There was a fellow employed at the mill who had a background as a cartoonist with a Winnipeg newspaper. He drew a cartoon that was printed up as a postcard and was kept by a lot of people. It showed a large group of easily recognizable people standing at the gates to the mental hospital at Essondale. Facing out is the keeper in a goatee, a black suit, and a tall stovepipe hat. Beside him is Tom Ryan, the superintendent.

The balloons say:

> Keeper: What drove you guys crazy?
> Tom Ryan: All those guys out there.
> Group (in chorus): TOM RYAN!

That kind of atmosphere pervades most happy plants. (When the plant is not happy, the crew draws unflattering caricatures of the boss on the plant walls.) I have worked for nine different employers, counting myself, and there wasn't one place where there weren't a few memorable people who made life enjoyable.

— * * * —

In my first years at Fraser Mills, we were still working a forty-eight hour week. Fitting forty-eight hours into a night shift and having any kind of social life was difficult indeed. We worked four nine-and-a-half-hour shifts and one ten-hour shift, which was done on Friday night.

On Fridays in the summer, we went to work in the daylight and came home in the daylight — 5 PM to 3:30 AM (1/2 hour for lunch), then an hour to get home, have a bite to eat, read the newspaper and unwind — then to bed at 6 AM, if we went at all. What is more, there was no swing shift. "Getting-on-days," as it was called, came as the result of your opposite number on days either dying or being moved to another job.

This mixing and matching of shifts often led to some domestic duplicity, a condition that had already been invented.

As life went on, there were little instructional episodes that helped me live a satisfying life later on. Others might say it was the birthplace of some long-standing prejudices, but if Wordsworth was right in claiming that "The Child is Father of the Man," I formed some long-lasting habits when I was eighteen years old. One of these notions has to do with pulling the trigger the minute you see which way things are going to go, and when it looks like the rest of the exercise is details and fine tuning. I'm afraid I didn't relate to General Montgomery's system where he wouldn't move a soldier until the last tube of toothpaste was in the supply truck. For better or worse I never did depart very far from these rules, although they were roundly debunked by a later generation which couldn't say yes or no until it had subjected every conceivable variation of a plan to microscopic analysis.

Harry Collier illustrates my point in a way. He was a good lumber grader and a man with a calm, self-assured manner. He was partnered with Stewart Alsbury, who was also one of Jack Munro's predecessors as president of the I.W.A. and whose brother, Tom Alsbury, was mayor of Vancouver for several years.

When Stewart was grading the flow of lumber down the chain, his face would often cloud over and we knew exactly what was going to happen next: he would suddenly run down the chain, find a piece he had graded minutes before, scratch out the old mark and give it another grade.

Harry would watch this with amusement and say, "Steve, mark 'em and forget 'em. If you know what you're doing, you'll be right most of the time." My application of this in later years was that it was better to have been mostly right and to have done something, than to have done nothing at all because you either couldn't make up your mind or were afraid of making a mistake.

— * * * —

An old nostalgic chestnut is to long for the days when men were men. The industry moved inexorably toward the use of more and more sophisticated machines, but a good peavey man was still fun to watch. It was just a dozen or so slick variations on the use of leverage, but it made heavy work look easy.

There was some remaining bull work in the mill at Fraser Mills, and for a while, Snap Stewart's grandson, Stu Windblad, had the dubious pleasure of performing it. For the benefit of the uninitiated, a gang saw is like an oversized bread slicer: large pieces of timber called "cants" or "flitches" were fed endways through numerous vertical saws so that several thinner boards came out the other side. The old Wickes gang, which was probably built before 1920, was sixty inches wide and each of the saws in the sash had to be tightened to the right tension with steel wedges. The farther in the wedge was driven, the more "strain" was put on the saw. This operation took a strong man with a five-pound hammer. When the machine was set up to make one-inch thick material, there were 51 saws, and Stu furiously hammered 51 wedges for half an hour without stopping because he had to do it between shifts so that there would be no lost production.

Naturally, someone perfected a system of straining gang saws with hydraulic oil pressure before gang saws were entirely replaced by circular double-arbor gang edgers and by quad band saws.

There was another bit of bull work that was exhausting for the operator, but comical to watch. Joe Bouthot ran an ancient vertical resaw that resawed slabs. The one controllable thing about a slab is that it has one flat side and, if it is laid with this face down and resawed with a horizontal resaw, the operation is simple. But stand a slab on its edge, and it is quite a different matter. When Joe resawed a slab he had to apply feed roll pressure from the side to hold the piece upright. There was a long lever that he pushed on to keep the rolls against the wood, and as the piece was fed in and the rolls had to climb over the knots and bumps of the rough surface, the motion was transferred back into the lever, and Joe would have to hang on for dear life.

— * * * —

One of my memories from when I was a kid is of waiting for my dad to come home so that I could rummage in his lunch box to see if there was a leftover jelly sandwich or a piece of cake or perhaps some candy bought from the candy lady, about whom there will be more later.

One night, before I could zero in, he beat me to it. When he came in the door he said, "Go look in my lunch box." I went, and what I found in there was a borer, which is a big, white wood-eating worm with accordion-pleated sides from its black head to its tail. It was more than five inches long and would have had to make a hole half-an-inch in diameter to make room for itself as it ate its way through the timber. I looked at it and looked at it, trying

to convince myself it was food.

Then, finally, I realized that I had been tricked. (These grubs are the larvae of the Long-Horn beetle. Malcolm Kerr said the worm in my father's lunch bucket was a runt. He has seen them larger than this).

— * * * —

Getting back to the candy lady, she was a middle-aged lady, considerably the worse for wear, who would come to the mill once a week with several large cardboard boxes packed full with small paper bags of candy. This poor soul would trudge all that long way up from the mill gate, staggering under the weight of her load. She would

put the first two boxes down at the time office and then go back for another load. Many times I volunteered to give her a hand, but most of the men left her to deal with her problem on her own. I can confirm that these boxes were a heavy load, even for a working man.

Once inside the mill, she made the rounds, selling a wide variety of sweets at ten cents a bag. Her assortment was consistent enough that people asked for particular kinds by name and when my sister and I were kids, we begged our father to buy our favourites.

Apparently the candy lady did this every day at different mills, and the rumour went around that she owned real estate and was well-to-do. What inner drive compelled her to torture herself this way is hard to imagine. We were coming out of the depression and there were still a few people who could see ghosts. It is possible that her income from peddling candy was handsome compared to the grim alternatives of the previous eight or ten years.

At about this time, a fellow started coming around with hot meat pies — they were both the forerunners of the coffee truck.

— * * * —

When I was a kid at the mill, the grading classes were held in a room above the company store. Learning to grade took a prominent place in our activities because it was, first of all, a steppingstone to a better job, since you couldn't move up if you didn't understand the product; second, there were examinations, and good marks provided you with credentials in the form of different classes of "grading ticket"; third, the winners of examinations got cash prizes from the company; and fourth, there was a provincial competition, and to win this was heavy stuff. The best I could ever do in the provincials was to come third two years in a row.

When I was on night shift, I went to these classes in the afternoon, then went to the home of an old family friend, Mrs. Tyerman, for supper before going off to work. She would feed me two helpings of rabbit pie with a thick crust and gravy and vegetables, and other hearty meat-and-potato-type things, with pie for dessert. I would waddle down to the plant feeling like I'd swallowed a ten pound weight.

— * * * —

A new kid from Quebec arrived to work on our shift.

He came with a lunch bucket that was crammed absolutely full of food. He would eat about half of it and throw the rest away. He continued to do this for several nights. When we asked why, he explained that he was boarding and his landlady was putting up lunches of this size without being asked. He said, "Maybe I'll feel real hungry some night so I'm not going to tell her and spoil a good thing."

— * * * —

Snoose and chewing tobacco were at one time synonymous with sawmilling. The mill office housed the foreman, the shortage clerks and a few transient tallymen who had in-and-out business. The waste paper baskets doubled as spittoons and it was your tough luck if you

threw away a piece of paper that you later wanted to

retrieve.

A lot of old lumbermen would rotate in their graves if they knew that the "in thing" these days is to make microwave popcorn in the mill office.

— * * * —

Occasionally a hardwood log would make a trip all the way from the logging camp to the mill. These were weeds as far as the loggers were concerned — usually maple or alder — but when the sawyer sawed the log up just to get it out of the way, you'd have thought it was rosewood. It would usually get out of the mill as far as the green chain, but as it moved down the chain, it would gradually disappear a piece at a time until it was all gone. None of it ever got pulled onto a load because everybody pinched a piece and put it away to dry.

I heard a variation of this story much later, but perhaps it was a bit more dramatic. While Bay Lumber was still located on False Creek, it became part of Whonnock Lumber. This brought it under the wing of Bill Sauder. Bill had a venerable cherry tree on his property which for some reason had to be removed. He was so attached to this tree that he had the trunk saved as a log. It was his intention to have it sawed into lumber so that whatever was made from the lumber would be a conversation piece.

Everything went according to plan and the log got up onto the log deck in the mill. Unfortunately, not everyone was told the whole story. "Someone," oh spare his name, saw it as just another worthless log and did the logical thing. He sent it directly to the chipper, where it soon became a thin layer of off-species pulp chips in the corner of a chip barge.

— * * * —

In the 1930's, pulp chips were not yet a revenue producing item. Lath, broomhandle stock, and shingle bands were about the extent of the salvage. Most mills had a refuse burner to dispose of slabs and other waste. These were later known as beehive burners, but that name demeaned the structure that performed this function at Fraser Mills. It was a huge cylinder with straight vertical sides. It was metal sheathed on the outside and lined with fire brick on the inside. The burner was then capped with a dome of steel ribs and mesh which limited the size of burning particles that could escape to the atmosphere on the swirling updraft. The whole structure was a hundred and eighty-five feet high.

Every winter there was an extended shutdown at Christmas for major repairs. One of the jobs was to rebrick any burned-out areas of the interior of the burner. Being young and money-hungry, I always tried to get hired on by the contractors or the mill maintenance crews for this work. John Hess did the burner repairs and I remember my first assignment working on scaffolding high inside the burner, hauling up loads of bricks with a winch and distributing them to the bricklayers. Some people may be born to work at great heights, but I kept feeling that at any time I might take a quick unscheduled trip back down to solid ground. For the first day or so, I literally left my fingerprints pressed into anything solid enough to hang onto. Besides that, it was colder than charity inside the burner because it was designed to draw a draft. But the money kept me interested and the butterflies eventually calmed down.

Harold Schiefke said that he and Ernie Lillos volunteered to work high in the air on the outside of the burner, replacing the screen on the dome. When they finished, the contractor had disappeared and they had to track him down in Vancouver to get their pay.

The wood scaffolding was left inside the burner when the job was done and became the first fire of the new season.

— * * * —

There had been a mill on the Fraser Mills site since before the turn of the century, and when a mill gets remodelled and added to it becomes a mystery story to be read and interpreted by misty-eyed veterans.

The braces and platforms and bolt holes and lean-tos and brackets could be compared to our creaking tax system, which seems to consist entirely of patches, loopholes and patched loopholes, to say nothing of amendments and addenda. To follow that analogy further, the country needs to do with its tax system what they did at Fraser Mills — tear it down and start all over.

The old mill was constructed entirely of timber, and whenever there were repairs to be made, any of a number of Scandinavian millwrights would remove old timbers and replace them with new ones. When they had to bore holes, they used long ship's augers which they dipped in grease to ease the periodic withdrawals necessary to clear out the shavings. These fellows could shape and size a timber using nothing more than a slick, which is a four-inch-wide chisel, an adze, and occasionally an axe. Cutting to length and notching was done with a crosscut saw which, when sharpened by an expert, could almost match a chain saw for speed. When they put the big pieces together, they fit like the joints in a jewel box.

— * * * —

The carpenters were good, but they weren't perfect. Most coast mill yards these days are made of blacktop over river sand fill. In the early days, the mill yard consisted of plank roadways that were built on timbers and mud sills or on pilings and caps, with the whole structure being six or eight feet above the swampy ground.

One year, the mill undertook the ambitious project of planking over several acres in a solid planked area. The millwright foreman who was in charge of the project started it from the opposite edges, laying sills, timbers and planks as they went. When they got to the middle, his calculations were twelve inches out and there had to be a little ramp built to overcome the discrepancy.

The foreman was fired, but since he was an old-timer with a pretty good batting average, they forgave him and he came back to work for many more years.

— * * * —

I remarked earlier on the social aspects of working in industry. One of the best experiences of my young life was to be assigned to Len Stone's spare list in the shipping department. Len had a labour pool and in the mornings he would get calls from the various departments for men to fill in for absentees. This was a great job because you never knew from one day to the next where you would be working. It added interest to the day and gave a fellow a more rounded exposure to a variety of people and the things that were going on around the plant. Of course there are still people who become quite upset if they are moved from the one job they do day in and day out.

— * * * —

The camaraderie and team spirit occasionally worked against the best interests of the company, but, heck, what are friends for? One of the planermen set his machine up on the wrong pattern and the run was well underway when someone noticed it. The anonymous "they" quietly threw all the off-pattern material into a conveyor that took it directly into the burner, the best place to put incriminating evidence.

— * * * —

When I got to the mill, a lot of the power was still being transmitted through shafts and belts — line shafts, flat belts and friction drives for power take-offs. A line shaft was a long rotating shaft driven from the main power source which would be a steam engine. This shaft would have pulleys of various sizes located at appropriate places along its length. These power take-offs drove various machines with long flat belts that danced and flapped with great activity and gave the place an added dimension of busy-ness.

The friction drives could act as a clutch or a corner drive or, in the case of circular friction discs, they could produce variable speeds by running the pinion at the slower moving centre or out on the faster moving rim, like the action on a record turntable. The belt man was a major player in the mill staff — his stuff made the machines go. In time, all of this was replaced by individual electric motors with reduction gears at each machine.

— * * * —

Out at the eastern edge of the mill property, behind the veneer plant, there were three distinct ethnic communities, namely Chinatown, Jap Town (forgive me), and Hindu Town (forgive me a hundred times). These last two terms were of course in bad taste because the Japanese objected to the word "Jap" and the East Indians were Sikhs and felt the same about Hindus then as they do today. As my friend Sawarne says, "You dumb white guys didn't know any better."

These places contained boarding houses for bachelors and for a few married people. They had common cookhouse and mess hall eating arrangements. There were also stores which provided ethnic foods and other things that the inhabitants preferred.

Chinatown was the best known of the three places because of a store, beer and booze joint, and gambling establishment run by Lim Sing. On paydays Lim Sing would cash pay cheques, a deadly arrangement which the company finally asked him to stop. People would hoist a couple, get into card games and, if they lost, would not bring home enough money to keep their family going until the next payday.

Lim Sing operated a keno-like game where you blacked out ten of eighty Chinese number characters printed on a square of rice paper. A similar "ticket" had twenty holes punched in it that corresponded to numbers drawn by chance as in a lottery. By laying the punched-out master ticket over your own ticket, your marked numbers would show through. If you got all ten right, you won $1000. Agents made the rounds of every job in the mill and most people bought tickets.

Back at the gambling room, a fellow would come around, beating on a dishpan and calling, "Open ticket!

Open ticket!" Inexplicably, this meant just the opposite — that ticket sales were closed and the next draw was about to happen."

Harold Schiefke tells of sitting in the old Edison Theatre in New Westminster when a J. Arthur Rank movie was about to start. As the muscular man beat the gong, Harold couldn't resist. He called out, "Open ticket!" and there were enough people there in the know to draw a big laugh.

The Japanese also had their own town. They lived together in their accustomed Japanese style, sharing communal baths and preparaing Japanese food. Most of the Japanese worked at the boiler house where they drew and wheeled out ashes from the pits under the fires.

Rosie Watanabe was the foreman of the boiler house crews and a Japanese bull gang that did the general outside maintenance for the plant. Rosie barked orders as though he was running a chain gang, but no one seemed to resent it.

When Pearl Harbor was attacked, the company reacted in the same way that most non-Japanese people did. Harold Schiefke and several of his cronies were playing 10 cent poker in the clubhouse when Maurice Ryan burst in and said, "They've bombed Pearl Harbor. Get the firemen out and go down and keep watch on the Japanese boiler house crew." They had a guard of sorts on all the Japanese workers for two or three days before things settled down. Shortly after that, the Japanese were settled in circumstances that we would all have on our consciences for a long while.

The East Indian group had their own community which included a temple. When Sikhs go into their temple, they remove their shoes. Two daring and irreverent young men once nailed down several pairs of sandals on the front porch of the temple and then ran for their lives.

The Sikhs at Fraser Mills still observed the ritual of cremating their dead on a funeral pyre, though they were later discouraged from doing this. It was done down at the edge of the river. I must admit that I never did see this, but young friends of mine who lived in the townsite used to watch in fascination from a distance until they were shooed away.

Some of the East Indians kept cows. One of these was the day shift foreman of the green chains and the remanufacturing resaws and edgers. (Yes, "white" employees reported to a Sikh foreman in the late 1930's at Fraser Mills. Who said we were rednecked racists?)

One night Shorty (Narain) sent over to the mill for someone to help deliver a calf that was breeched. We had a Finlander on the chain who could hardly speak English and what he did say, he growled in a way that was impossible to understand. The foreman volunteered him and he came back hours later, not having washed, to describe incredulously how he had been into the unfortunate beast up to his shoulder, trying to turn the calf. The communication at that scene must have been bizarre.

— * * * —

I had only been at the mill for a few months when the war in Europe broke out, although I did not join the navy until 1942. I turned seventeen in January of 1940.

Right after December 7, 1941 (the day the Japanese bombed Pearl Harbor), there was a welter of rumours, most of them false, about Japanese planes approaching the coast. One night there was one of these scares, and the lights were turned off while we were working. We had to find our way out of the plant in the dark. I remember driving home in my 1929 Pontiac roadster with my lights turned off. After that, we taped our headlights

so that just a small horizontal slit of light showed through.

— * * * —

J.R. Murray was the general manager of the company and Frank Jordan was the superintendent of the plywood plant and door factory.

J.R. Murray was also the president of the New Westminster Adanacs Lacrosse Club and Frank Jordan was its general manager.

At the end of the day shift, therefore, it was not surprising to see practically the whole Adanacs team streaming out of the door factory to take the long walk to the gate with the rest of us — Gordie Saunders, Bob Lee, Jimmy, Johnny and Ralph Douglas, Punch Thompson, Dutch Haddon, Eddie Johnston, Brud Matheson and many more on that 1939 Mann Cup championship team. It drew a lot of comments at the time because jobs were hard to get, but in a year they all had much different jobs at which they distinguished themselves and for which they drew nothing but glowing praise.

— * * * —

I worked at Fraser Mills for three years, then did three and a half years in the navy before coming back to the mill for another go at it. I remember going in to see Maurice Ryan to ask his advice on what to do at this point in my life. Being an ex-serviceman, I could have gone back to school, picked up what I was missing for high school graduation, and gone on to university at the government's expense. Right or wrong, it seemed at the time like an exercise in time wasting since I had a wife and a job and something of a start at a trade, if not a profession. Besides, there wasn't anything in particular I was dying to be.

Maurice said that he thought that I could do worse than stay in the industry. "As far as this company goes, we have enough timber to last our lifetimes," he told me, "and after that we won't have to worry." Within six or seven years Crown Zellerbach had bought the company, Maurice had gone on to other things, and I had made the second of many moves myself. No gold watch for me.

— * * * —

iv) Sailing ships at Fraser Mills, early 1900's.

v) Sailing ships and skids at Fraser Mills, early 1900's.

vi) Fraser Mills gate and townsite 1923 — the year I was born there.

vii) Horses and teamsters, about 1922.

viii) "When the sawyer turned a particularly large log, the thump when it dropped onto its flatted side would feel like a minor earthquake in the townsite." To be honest, this was taken at Dollarton Mill, but you get the idea.

ix) 94' long timber, probably 16 x 16, 1940.

x) Cargo accumulating on docks at Fraser Mills.

xi) The "back end" of the mill at Fraser Mills. Everything about the old crane and its track creaked and crackled as it travelled.

xii) A nice log heading into the mill at Fraser Mills.

xiii) Air dry piles. The roof boards and the overhang built into the pile shed the rain.

Chapter 3

The Man in the Middle

In 1947 I was asked to join the Pacific Lumber Inspection Bureau. In those times and for the level of work I had been doing, getting onto the PLIB was considered a prestigious move. I left the employ of the company, although I continued to do my new job at the Fraser Mills plant.

The PLIB is a venerable institution that performs the function of an independent grading and tallying service which assures offshore customers that they are getting the quality and specifications that they ordered.

Being in the middle does not necessarily make for a fully rounded man. A few of the inspectors had the reputation of being hair splitters. In an effort to avoid compromising their impartiality, they went by the book to a degree that ruled out the application of common sense. Lumber being a natural product, its segregation into categories of quality (grades) required a broad mind and some judgement. The expression, "suitable for the intended purpose of the grade" was the judgement element that measured how sensible a grader you were.

I have seen situations where inspectors rejected thousands of feet of lumber for imperfections that were superficial. I once stuck my neck out and began accepting material on the basis of my own evaluation of the seriousness of a particular condition. Nothing happened; the world didn't end and years later the manager of that plant said that Jess Archer and I were the only inspectors who ever gave the company a break.

Of course, most of the bureau men were extremely capable, working a dozen different orders at a time, applying different lengths to different orders, working different percentages of grades on different orders at the same time, reaching maximum allowable amounts on orders and applying the overflow to new orders, and all the while being sure that the right markings and lot numbers were being applied to everything. And they did this while the lumber was flowing at high speed out of production machines that weren't going to stop for anybody.

The clearest head I ever knew was Bill Lereaux. He was a human calculator and it is hard to see how he could do what he did without a machine and a legman to chase information for him.

Other bureau men worked at a more leisurely pace, reinspecting material that had come pre-graded from the mill.

George Gunn was a fine old gentleman who worked the last part of his career with the understanding of supervisors who tolerated a unique partnership. George's eyesight had almost completely gone, but since he was an inspector, a major accommodation had to be made for this. He had a right-hand man in a Finlander named Salem who became George's eyes, ears, and virtual life support.

A typical day would see George inspecting Douglas fir 6x12 and wider, up to 40′ long, for export to Australia. In addition to Salem there would be two men with peaveys to turn the pieces. Salem had a 10′ stick with which he walked off the lengths and a number of paint daubers with which he put a colour code on the ends of the pieces to identify the grades.

Standing to the side, you could hear Salem call, "Sixteen" (the width), "thirty-six" (the length), "Select" (the grade), and from this information George would make a large tally stroke on a sheet of paper. At the end of the shift, George, with the aid of a magnifying glass, would recopy his day's tally onto a proper tally sheet, and no tally sheet ever looked neater.

George knew how to take a drink and when a ship would come in from the West Indies, it was usually pretty easy to find a bottle of bootleg rum. Late in the afternoons on those days, George would get mellow and Salem would have to do the whole job himself.

— * * * —

I put in a lot of years in the industry and never lost any time from a serious injury. Looking at workers from the management side, I have made my share of scathing remarks about people with sore backs. Sore backs being the invisible wounds that they are, everyone who has one is suspect because of the odd guy who has been caught helping his brother-in-law move his furniture, or is found on a sunny afternoon taking in hay while on back-related compensation.

I had a sore back once and it honed my insistence that machinery be designed and installed to suit the operator as much as possible. I was working on an opposite shift from Manny Bouthot, inspecting lumber on a cargo chain. Manny stood about 5′4″ and I was 6′1″. Manny, being the industrious little man that he was, built up the floor so he could see what was going on. I, on the

other hand, worked in a crouch for the best part of two weeks. We did the sensible thing in the end and built a removable floor, but not before I was nearly crippled. It is a question every supervisor would be wise to ask the employees: "Are you physically comfortable at your work station?".

My partner on the bureau for a year or so was a puzzle. When we worked together, he remarked again and again, "I've got a bad heart and I have to be pretty careful." He was never without those little vials of nitro glycerine to break and sniff in an emergency, though I never saw him do it.

He worked for another thirty-five years after we worked together and he lived into his eighties.

— * * * —

One of the enjoyable parts of PLIB work was being assigned to transient work; that is, getting a call from the supervisor the evening before, then heading out to unknown territory in the morning. Some of these assignments were pure adventure.

One of the many mills along False Creek was Bay Lumber. This mill was operated by John Fiegl, a portly and imperious German gentleman whose dour mien and quick temper were well-known in the industry.

Over the years I have been able to get along with just about everybody, but John found the right way to get me ruffled. He was bringing U.K. railroad ties (quaintly called "sleepers" by the British) down from a bush mill in Lynn Valley. My order was for hemlock but the species on one of the trucks was fir. Since U.K. ties were cut to a special size, the fir was going to beg for a home for a while, and this presented John with a dilemma. He thought he would foist his dilemma onto me.

He announced that fir was better wood than hemlock any day so I should just put them in the order and say nothing. I refused and we all got red-faced and exercised. The next morning I was on my way to another mill.

I had a friend who worked for John. Before Bay Lumber joined Seaboard Sales, John tried his hand at selling East Coast U.S. cargo direct to the New York wholesalers. He got caught in a market slump that caused his customer to cancel his order after the ship had sailed and was, in fact, going through the Panama Canal. This got the wires buzzing, but John eventually had to unload and store his lumber on the Brooklyn docks and pay demurrage until he could sell it.

John was furious, and at one point when he got his New York customer on the phone, he called his secretary and said in his heavy German accent, "Elsinore. Take ze uzzer phone and take down effresing vot he says."

This same Eleanor, which was her real name, played a part in a later encounter with Feigl but I will tell it now.

Eleanor was subtly letting it be known around the industry that she could be persuaded to move. This information reached the company where I was working and we hired her.

I was buying lumber for a wholesale company by this time but I had never been able to get to first base with Feigl. One day, I was making what in the past would have been one of my futile visits when, instead of being brushed off, I was ushered into the inner sanctum. John asked me to sit down, heard out my pitch, and asked me many encouraging questions, but he suddenly stood up, strode to the door and, opening it, growled, "I vill tell you somesing. I vill do no business mit beoble who steal my help."

I was being kicked out!

— * * * —

One night Bob Walden called me and gave me an assignment which was to go to Chapell Lumber on the North Shore waterfront below where the Second Narrows Bridge now stands.

This was a routine inspection, except that Bob said, "I want you to take a small tri-square with you." He added, "Chapell's carriage is pretty sloppy and a lot of his squares come out diamond shaped. Don't take anything that is more than an eighth of an inch out of square." Well, the rest of the story was that Chapell was late getting his order out. The necessary logs had arrived just the night before and the ship was in the harbour waiting.

When I set up to inspect the material, which was rough hemlock 8x8, it developed that John Chapell himself and a helper were going to turn the stock for me, handling the pieces from one slingload to another. Green hemlock is as heavy as iron. Backbreaking is the only word that could describe what they were doing. They started working furiously and kept going until we had gone through the whole parcel, most of which was coming hot out of the mill as we worked.

The few times I tried to get my little square in they would impatiently say, "That one's O.K." Since I didn't want to lose a hand, I took their word for it.

Reinspection was another part of transient work. Being the arbiter in a dispute about quality was always a bit traumatic — if not for the inspector, then surely for either the buyer or the seller.

If the material was grossly off-grade or completely on-grade, someone's face would be red and that was the end of it. However, if it was only a little bit off-grade but technically beyond the allowable limits of variation in quality, the loser had to settle for the off-grade material and pick up the reinspection costs as well, and he usually felt that you had crucified him.

— * * * —

Chapter 4

A Little Travelling Music, Please

After awhile I began to feel that there should be more to the lumber business than nuts, bolts and sawdust.

On a whim, I answered a classified ad for a lumber buyer for a wholesale lumber company. This was in 1949.

Since wholesaling is a transient business, or at least was when there were many more mills and much more travelling and personal visiting was done, it would be easy to lapse into a singsong litany of then-I-did-this-and-then-I-did-thats. Still, barnstorming is exactly what I did, and though wholesaling didn't make me rich, it gave me a chance to meet interesting people, see the country, visit many mills and observe the way things were done by a lot of hard-fisted, practical people. Wholesaling lumber involves buying from a mill and selling to a customer without ever seeing the stock. It is a paper transaction and a commission is earned, but it is not a commission in the sense of a broker's commission. The supplying mill sells it to you at his price and doesn't care who you sell it to or for how much, as long as you pay your bill promptly. This means that as soon as the mill has invoiced you for the material, it is yours, wherever it may be, and how you get it to a place where you can sell it at a profit is nobody's business but your own. This was a part of the lumber industry about which I knew very little, but I went for an interview and made the acquaintance of one of the fairest and bravest men I ever met.

Jack Webster had had an interesting career in South America. If he had not been falsely charged and eventually jailed in 1936 for supposedly supplying inferior war materiel to Bolivia for their ill-fated Chaco War with Paraguay, he probably would have stayed there. The circumstances leading to Webbie's plight did not seem to have much substance since he was buying from good American suppliers like Remington Arms and Dupont. It seems more likely that the Bolivians came down into the steaming jungle from the high elevation and pure air of their own country and tangled with some nasties who poisoned their water and hid in the trees and shot at them.

The wooly headed British bureaucracy took a year to get Webbie out. The Americans were more direct and the Standard Oil man was out in a week.

When Webbie got out, he went to Chile, but there was student unrest there. When the army pulled its artillery up in front of the university and started shooting holes in the buildings, he thought it was time to leave. He came to Vancouver in about 1948 and bought into the George L. McNichol Company, whose president at that time was Frank Bernard.

Though that company's business was built around cork products, and they were a general import and export company, they had a lumber division. Jack was a man who would buy and sell a shipload of tallow and work it for half of one percent commission, so it wasn't surprising that one of their early deals was to buy a war surplus LST, fill it with lumber and sell it, ship and all, to a customer in the Argentine.

Shortly after I joined McNichol, Jack left to form his own company, J.W. Webster Co., and I went with him. He operated this company until he died.

— * * * —

Jack worked full-time for many years with the inconvenience of a colostomy. This was the result of an unsuccessful attempt by a South American doctor to repair a First World War abdominal wound. Travelling delays were murder for him and the limits on his diet kept him painfully thin, but he pulled it off with a smile. As long as he was comfortable, he was an absolutely charming man.

He was patient, considerate and calm. Being this way allows you to make an enormous impression when you step out of character. We had a no-account working for us for a short time, to whom Webbie, by his nature, gave every chance to measure up and change his ways. One day Webbie discovered that the guy had done something dishonest and he yelled for all to hear, "Get out! Get out! If I never see you again it will be too soon." The guy ran to the door in genuine terror, even though he was twice Webbie's size.

If we were on the road in a small town such as Princeton, and we had time to kill in an evening, we would shoot pool in the local pool hall and take our turn with the regulars. (Jack also played billiards and snooker as a member at Vancouver's very exclusive Terminal City Club.)

— * * * —

If, at five-year intervals, you listed the established mills in B.C., there would always be a significant number of add-ons and drop-offs, so I suppose any personal account would be at best a snapshot picture of a particular day, rather than a comprehensive history. For instance, the mills around False Creek in 1949 were, starting at the Granville Street Bridge: Fairview Sawmills, Robertson and Hackett, Sweeny Cooperage, Marine Lumber, Nalos Lumber, Bay Lumber, Alberta Lumber, E.L. Sauder, Sitka Spruce Lumber, A&M Sawmills, Cedar Cove Lumber, B.C. Fir and Cedar, and Giroday Sawmills on Granville Island, and I may have missed a couple. Many of them were steam powered, and a pall of smoke from their boiler house stacks and their refuse burners hung over the whole area as a result.

The Sitka Spruce mill was interesting. It had become one of the B.C. Forest Products mills when that company was formed. It disappeared in a spectacular fire in 1959. This mill specialized in cutting huge Sitka Spruce logs at a time before the "square log" remanufacturing cants for the Japanese market had come on the scene. The main product was, of course, clear lumber, and some of it was still graded under airplane spruce grades, even though it may have been used for ladder rails and other high quality products. The grader used an ice pick to pull a sliver to measure the slope of grain, since straight grain was the key to the quality of high grade spruce clears. As large as they were, each log was broken down into specific grades and sizes, and the sawyer had to extract the last dollar of value. It was a veritable feast.

Because these logs were so big and often had flared butts, they did not come into the mill by jack ladder. They were floated into a well and hoisted up to the log deck with a cable parbuckle. Many logs were too big to bring into the mill whole, so they were first split into halves and quarters in the river by a splitter saw designed for the purpose.

— * * * —

The size of the head saw and the carriage used to determine how large a log a mill could cut.

When pulp chips became a major source of revenue, most mills installed debarkers. Bark is not acceptable in pulp chips because of its different fiber and because it is difficult to bleach. For ecomonic reasons, most mills installed a barker that would only handle their "average" sized log, and so many logs that would have formerly been sawn without difficulty became "oversized".

Mike LePage has a log splitting business that his father started in 1946. Because of the circumstances mentioned above, it is going stronger than ever, with logs being towed up and down the river and out of Burrard Inlet to two different saw locations where they split and quarter ten to thirty logs per day (40M FBM), the logs ranging from five feet to fourteen feet in diameter.

— * * * —

On the North Shore, Larry Lytle of L & K Lumber was good to me when he didn't have to be — there was quite a fraternity of wholesale companies all trying to work the same wood. L & K had hit on an interesting specialty in long length joists and purlins — 2x12 and 2x14 up to 28′ long — and did very well at it until Larry died tragically when the plane he was flying crashed on Redonda Island in 1959.

Also along the North Shore was North Shore Lumber.

Bill Fairburn was there then. Bill is a good friend, a good lumberman and an enemy of pretense. When the industry was in the throes of "getting big" and new titles were being coined almost by the hour, Bill was scoffing and holding the line on staff. Not surprisingly, he was later a part of a model organization, McDonald Cedar, where a minimum number of people achieved a maximum amount of positive result.

— * * * —

There is a North Shore Lumber story told by Ron Fielding on himself.

When Ron was working there as a millwright, he and a partner were in the mill on a Saturday morning to do some structural timber changes. They found that a timber that they were going to replace was an odd size and the only piece they had for a replacement was too big. After a bit of head scratching, a light came on. The mill had a timber sizer (planer) and they reasoned that there surely couldn't be that much mystery in setting it up to plane their 12x12 timber down to the right size. Well, many a lumberman before them has been forced to conclude that planer work is a black art. They set their size and started to feed the timber in. It went along fine until it was halfway through, then it stopped feeding and no amount of coaxing and fiddling and adjusting could get it started again.

Realizing that they were now in a potentially embarrassing position, since most planermen consider their planers to be their own private property, it was obviously time for bright idea number 2: timber sizers look simple so a simple solution would seem to be to somehow give the timber a modest boost to start things moving again. The sizer stood by itself in a small shed in the mill yard, away from the main building. The company had a truck, so why not, they thought, just put a chain around the protruding end of the timber, hook on with the truck, and give it a pull. They did this, again with no results, but while they were razzing the truck, backing up, running ahead and jerking on the chain, the planerman drove

into the plant. As they used to say in more sheltered times, "I leave the rest to the reader's imagination."

— * * * —

It's not so long ago that there were small portable mills here and there around the valley cutting little patches of private timber. A fellow named Fred Bernard had such a mill on Crescent Beach road. We had given him an order for some mine timbers and he produced them from second growth hemlock. He loaded out a textbook flat car that looked like a package from Birks. The railroad hauled it as far as Abbotsford before they weighed it. It was so heavy that they wouldn't even haul it back to the team track where it was loaded. We eventually made two regular-weight carloads out of the one original load. Lead would have been lighter. You learn a little every day.

— * * * —

I hit it off with Bill Robinson at the old Royal City Sawmills. This was one of the first mills to give a woman a chance to use her brains in a male community. Dorothy Ritchie was a very capable shipper there and later at Capilano Lumber, and no one even thought to question the fact that she wasn't "one of us."

Two of the pioneer names of the lumber industry reappeared around 1950. Murray McNair and Fred McVicar had a mill called Argyle Lumber at the foot of Argyle Street below Marine Drive. Straightforward and honest, they put a considerable amount of business my way and let me get a feel for how business was done.

Every wholesaler has a mill where he tries his hardest but never seems to put any deals together. Sol Seidler and Morris Miller ran Seymour Sawmills but they sure didn't run it for me. I just couldn't find the handle.

— * * * —

Newton Cameron had a mill at Leechtown on Vancouver Island. The fir where he was logging grew tall and straight like giant candles. There was a timber item of another day called "transmission crossarms." These were

4x12's that ranged in length from 28′ to 34′. They reached between two power poles (in pairs) and the high voltage power lines hung from them on insulators. The grade of the material was very exacting because of the critical function the crossarms performed, but with the timber Cameron had, he cut them as though they were 2x4's.

— * * * —

A West Coast kid was never far from a sawmill. I must mention the old Dominion Mills, which became part of the Canadian White Pine division of MacMillan Bloedel, because it reminds me of one of these indelible bits of nonsense from my childhood. We occasionally visited friends who lived at Marine Drive and Kerr Road, just above the mill. Tom Holmes was a PLIB man and they also had a corner store there. Tom was once robbed and held hostage by William Bagley, the famous B.C. bank robber and murderer.

From the Marpole interurban tram, you could see the head spool of a conveyor that was discharging some sort of wet pulverized waste. It always looked to me like it was doing something nasty, the way it oozed out over the end.

— * * * —

There was a mill in Port Kells that had been called Wades Mill but which later became Port Kells Sawmill. Mrs. Wade, the former owner's widow, had married a fellow named Harold Morrison who now ran the mill. He talked to me a lot and always encouraged me to learn as much about sawmilling as I could for future use. One day I found out why.

Harold's wife had died and shortly after he asked me to come around to see him on a Saturday morning at his office. This old unpainted building, for some reason, had weathered a dark brown and it was as basic as it could get. Harold lived in modest quarters above the office. The reason for the invitation? He wanted me to take over the running of the operation. He would stay on as a father figure, to advise me, and I would gradually buy the place out of profits earned after I took over. Either I didn't realize the significance of what he was saying, or I thought I wanted to stay with wholesaling, but in any event, like an idiot, I declined. This mill is now the "A" mill of the S & R group.

— * * * —

Orion Bowman was operating his mill at Sardis. Mr. Bowman was a natural man. If I had called him "Nature Boy," a number of people would have nodded in agreement since there was a song with that title popular at the time. He was not young, but he walked around with no shirt, his pants held up by a piece of rope and, as often as not, was in his bare feet.

He creamed his logs — sold the best to log brokers and sawed the junk in his mill — and so he should have if that is what made him the most money. I had the audacity to question this policy one day and Bowman's daughter, Kay Mcleod, angrily let me have both barrels. If I didn't like the way they ran their business, I didn't need to call around. It's known as minding your own business.

There is no question that Bowman did things his own way. He would occasionally have the mill sawing Locust logs to supply someone with material to make insulator pegs for power pole crossarms.

— * * * —

Harry Pehota ran a mill at Popcum Siding called Coast Sawmills. I believe his partners were Bud Lando and the Pedens from Victoria.

When Toronto built its subway, they excavated from the top down, rather than tunnelling. They built a roadway over the top using 12x12 timbers, to allow the work to proceed underneath without disrupting traffic.

It was well-known that a certain wholesaler had been the successful bidder on a large amount of this timber. It was also known that he had sold it without having his purchase covered and he was scrambling to meet his commitment.

I walked into Harry's mill and saw some 12x12 on the timber deck. I made a complimentary remark about the quality and asked if he would supply me if I rounded up some business. He said, "That depends on the price you offer." I gave him a price indication and he said that if I could get an order at that price, I could put a few cars on.

Well, we did, and we sold them across the hall, so to speak, to the distressed wholesaler. When he saw the railroad bill of lading and realized that he had bought lumber produced at a mill where he had orders placed for the same material at a lower price, he was fit to be tied.

— * * * —

Jack Munsie ran Kettle Valley Sawmills near Princeton. I bought a few cars from him while he was still cutting big old growth Ponderosa pine into clears and shop grades. Every lumberman past or present deserved to see that quality of lumber so that he could take a sweet thought to his grave.

— * * * —

Rock Creek is just around the corner from nowhere. One night two of us pulled up at the Rock Creek Hotel and asked the Chinese clerk behind the desk for a night's lodging. The conversation went something like this:

"Do you have two rooms?"

"Only one room left."

"Does it have two beds?"

"How many people?"

"Two."

"Two man two lady?"

"No, two men."

"Two man, one bed. Five dollar."

Rock Creek in the summer is like the hot side of hell. What is more, our room was directly above the beer parlour. The regulars whooped and hollered the night away and finally said their goodbyes in the parking lot under our open window. We finally did get to sleep, but were soon wakened by the roar of what seemed like at least fifty logging trucks changing gears on the hill in front of the hotel.

— * * * —

For the benefit of anyone who has not been in the wholesale end of the industry, I should explain that there is a unique way of doing business that used to be very prevalent and is still done to some extent today. It is called "transiting," and it consists of rolling unsold carloads of lumber in the general direction of the major markets. This is done on the premise that having rolling inventory, with the available lengths, the grade specification and the price established, puts you in a position to make an easy sale to someone who wants what you have in a hurry.

Lists of this rolling stock are widely distributed and in a strong market, transiting works. However, in a weakening market, or one suddenly distressed by a presidential tummy ache or by some government financial jiggery pokery, a transit car quickly turns into a millstone. The expression, "having one at the bumper," means that you have advanced a car to the point where the options to divert it to a variety of places are about to run out. There were times when the heavy gamblers would have 500 unsold cars heading for the Minnesota Transfer at Minneapolis — with the thought occasionally crossing their minds, "Gee, what if they never sell."

A rule of thumb that is as good today as it was in 1951 is, "Don't fall in love with the lumber; fall in love with the margin you've got in your deal."

I saw some pine boards at Trautman and Garraway in Peachland that were so beautiful and well milled that the only question I had to ask myself was, "Should I load one width per car with some of each grade, or should I load one grade per car with some of each width." The utter idiocy of rolling something like this didn't occur to me in my eagerness to buy that wonderful lumber. The stock wound up on consignment in Cherry Lumber Corporation's sheds in Philadelphia where it eventually sold for an embarrassing loss.

— * * * —

I mentioned earlier that this book would not likely chronicle the lives and times of the rich and famous since I didn't get to play in that league. I did know a few guys whom I considered to be exemplary and one of these, Horace Simpson, bears mention because he had to live with two things that for some could have been handicaps. First, he was his father's son and Stan Simpson was a well-known achiever; also sons don't always build on their fathers' good foundations. Secondly, you'd have to be made of special stuff to make it through life with the name of Horace. He obviously had no trouble with either of these conditions because he took the company on to even bigger things before it was sold to Crown Zellerbach.

— * * * —

The time that I am talking about now is 1950 and to a guy who thinks that the old days were in 1935, 1950 is just yesterday. At one time you could say that there was a sawmill around every corner in the Interior and not exaggerate. In 1950 there were a lot fewer than there had been, but there were still people like Jim Churchill at Falkland who lived out the old cliché of doing it because it was there. He got a timber sale, logged it, hauled it to the mill, sawed it himself (he was a sawyer), planed the lumber (which was all 2"dimension), and sold it to guys like me. He didn't have anyone doing marketing, sales promotion and product development, and trendy euphemisms didn't garble his messages.

— * * * —

There was a road going from Falkland that, if

persevered with, would come out at Chase. It was called China Creek Road. Once, when I was following a lead, I went up the road ten miles or so and turned off on another wilderness road in blind faith. After a few more apprehensive miles, the view opened up and the road led into Shangri La. Before me was a lush green valley

bottom called Spruce Valley. There were cattle in the meadows, farm crops in the fields, a settlement of buildings, and of course a small sawmill, which was my reason for going there.

This was a Seventh Day Adventist community and the prominent person there was Ed Reiswig. There was a common dining room, a chapel, and private married and single quarters for the residents.

I was asked to stay for supper and we were served a shepherd's pie made of some kind of cereal in place of meat, and a fruit cordial to drink. Everyone was very pleasant and friendly and unexcited.

I wouldn't want to go back there for fear that it had turned into an open pit mine or perhaps had a highway running through the middle of it.

— * * * —

One day I went to the little mill at Malakwa that the Shellys from Prince George had built. When I went into the mill, the sawyer waved to me. We established where we had met before, but I won't tell.

The story we recalled was of how, at this other place, sometimes the first order of business in the morning was to saw up "those logs."

"Those logs" were the overnight deliveries of beachcomb logs that might still have some identifying timber marks on them. The beaches these pirates combed didn't have a lot of sand on them.

— * * * —

Brick McDonald and Jim Burr had a mill at Grindrod. They were both Coast men who had made the conversion to Interior lumbering painlessly. Brick and I worked out a specification for a high grade of fir plank that was used for hatchcovers on new freighters. We sold it to the U.S. Shipbuilding Corporation in Boston. One size, one length and a high grade — that broke a few rules — but Brick could do it with the timber he had.

There were a few mills around Ashcroft and Cache Creek, an area where the timber was practically dry on the stump. The lumber sawn from it would dry to a crisp within a few days in the blazing summer sun. Ashcroft and Hedley and Lytton and Lillooet get my vote for the most desperate lumber propositions that I ever saw.

Bill McQuarrie operated a mill at Hedley for several years and we were good friends.

I dropped in on him one afternoon and he said, "Hop in the pickup; I've got to go up to where they are logging before they knock off for the day." We took off on one of those typical dusty, bumpy trips on a twisting road up, up, and up the hill. Eventually things flattened out a bit and we were in a park-like setting with trees standing 40 to 50 feet apart.

I asked Bill how much farther we had to go and he

said, "This is it. We're here."

As meagre as Bill's logging prospects appeared to be, he would still cut an eight or ten foot long log off the butt of most trees and from these he sawed high value transmission crossarms. He sold these to Brick McDonald who had an arrangement with B.C. Crossarm in Surrey. Networks of connections make business happen.

There was a mill at Clinton that was bought by the old Fraser Mills group of Maurice Ryan and Bill Mackin. Harold Scheifke was the resident manager and resident anything else that needed a set of hands.

They ran their own sawmill and also bought rough lumber from small mills in the area. Harold told me that when he arrived, he inherited 200m of reject material, rejected by someone very picky. Harold processed it all in the normal manner, into normal grades and sizes, and sold it all for normal prices. Everyone dreams of a little money for jam.

— * * * —

From time to time my path crossed with that of a gentle little man named Julian Bass. Julian was a lawyer from Seattle, but he had various interests in the lumber business as well.

One of our meetings involved a junket to Blue River, B.C. and our objective was to look at a mill that was being offered for sale along with some timber.

Blue River was only a hamlet and this mill was out of town from there, so we were really in the sticks. The road from Kamloops did not yet go up there, so we went in on the "way freight." In other circumstances the way freight is known as the milk train. It was pulled by a single engine and it supplied empty box cars to mills and removed loaded box cars from mills along the line. This was more than a switching operation because it also had one passenger car and if you had the time you *could* get there from here. By this dawdling process, we were eventually dropped off at the mill.

The mill was shut down and had that white weathered look of an abandoned Interior operation. There wasn't a soul around to greet us and it was late afternoon. This was not the plan, but we must have had a premonition because somewhere along the line we had acquired a can of beans. We ate these cold, from the can, passing it back and forth until they were gone.

It was summer and that area can be like a furnace, so when night came we just slept on the floor of an old shack like a couple of well-to-do hoboes.

The following morning our local contact showed up.

He got the deisel power unit running and started up the single saw circular head rig. There were two or three small, dry old logs on the log deck and we loaded one of these onto the carriage. The saw was about halfway through the first cut when it suddenly protested, almost stalled, then tore the log free of the dogs and fired it right out the end of the mill.

We didn't buy and I never heard of the place again.

— * * * —

Two Mile Flat at Quesnel at one time accommodated twenty planing mills, all of which were buying from bushmills that were busy gnawing down the forest in small bites. Some of these mill operators were sharecroppers, so to speak, sawing timber owned or financed by the planing mill operators, while others were independent operators. The mills were small so-called portables, and most of them would only cut 10m FBM a day.

A Quesnel banker freely admitted that there were more business deals done in the beer parlours than there were in his bank. And it was frontier banking; they had to stay in close touch with everything that went on in the community since people were inclined to disappear into thin air. We got a call from this same banker one day. He had a carload of railroad ties to sell, this being the only thing he could salvage for his bank from a disillusioned entrepreneur's dreams.

And it was dream time. There were hundreds of small operators in the Cariboo and it was not unlike my father's Klondike days, with people needing just a little backing to get their Juggernaut in gear. The odds were against them, but it was a lively scene while it lasted.

The lumber they cut was brought to the planer mill where it was passed over a sorting chain for trimming and grading. The value settlement was made on this grade and tally and many of the mills were so chronically short of funds that they would try to get paid in cash, a truckload at a time. They always needed machinery parts and had other gnawing expenses. Most of them settled up by the week or month, but it was a put-it-on-the-bill-and-I'll-pay-you-next-Friday economy.

In a few years, even the Cariboo, a sort of last bastion of the small entrepreneur, would undergo the too familiar transformation into a limited game for a few large companies.

While sophistication and its accompanying lawyers' fees eventually became the only route to an airtight deal, the handshake served a lot of honourable lumber people well.

Occasionally someone would hit the scene who was not so honourable and this would give handshaking a bad name.

A square and sober fellow I knew was given an assignment in Australia. Since he knew that he could be gone for two or three years, he had to tidy up his personal affairs before he left. He owned some timber in the Cariboo and decided to have a so-called management company look after getting this timber logged and sold for him in his absence.

When he returned on his first leave about a year later, he checked to see how things were going. He found, to his dismay, that there were no trees left, but neither was there any money to show for them.

When he howled, "What have you done?" the response was, "You think you lost money. You should see how much we lost on the deal."

This same outfit, who were also lumber brokers, once financed a new power unit for the sawmill of a friend of mine. This was to be repaid over a period of time. They got in a financial squeeze themselves so they encouraged the mill to load out several cars of lumber for them all at once. Once they had the bills of lading and thus title to the material, they informed the mill operator that they were offsetting his machinery debt with lumber and that they were now all square. It broke the little mill.

On a trip to the west side of the Fraser to an area called Narcosli Creek, I had a different experience. The gumbo country road had dried hard in ruts and potholes in a snapshot picture of how it had looked during the spring breakup when, of course, it was mud. I was driving a 1947 Plymouth station wagon, and as I eased myself out over this road, the car lurched and bumped in every direction.

Occasionally I would hear and feel things scraping on the high centre of the road, but suddenly the ruts got really deep and I felt the pan hit with a grinding crunch. Upon inspection, it was clear that I had torn the plug partway out of the pan, and the oil was pouring out on the road.

I was nowhere, but amazingly, I was were within a few hundred feet of the only house within miles. This was an assembly of old rustic buildings on a typical stump ranch. As I approached the house, I noticed that the washing machine was on the front porch. This seemed ordinary enough under the circumstances, but it proved to be my salvation. This was a gas-operated washing machine and the lady of the house offered to sell me the couple of quarts of oil that she kept on hand for her Briggs and Stratton washing machine motor.

I got out and got under, stuffed some rag into the hole in the pan, and pounded the tattered plug over the rag to keep it in place. I put in the oil and started off. I thought I had it made, but I had a long way to go, and before I got back to Quesnel the engine was hot and I could hear the ominous sound of bearings knocking. I nursed the car into Quesnel, had the plug brazed back in, gave the engine a big drink of oil, and headed for the Coast with the engine knocking gently and my fingers crossed.

Just before Clinton there is a long steep hill that goes right down into the town. A mile or so before the brow of the hill, the motor started to sound like a jazz drummer's solo. Just as I got to the top of the hill, there was the loud bang of a con rod breaking loose and hammering around in the engine block, and that was it. I coasted down the hill into the waiting arms of a mechanic who needed the work, and went the rest of the way back to the Coast by bus. My insurance company agreed with my contention that the car had been in a collision and paid the claim.

I told that story to Ken Ross, who said, "When that happened to me, it was 14 below and I was in street clothes. I tried to plug the hole with some body putty I had in the car, but by the time I got at it, everything was so cold the putty wouldn't mould. I lit a fire under the engine to warm things up but I set the engine on fire." Somebody is always trying to go "one up" on you.

A trailer to that conversation was added by a wilderness veteran. He said that if you plug the hole any way you can, but can only salvage a quart or so of oil, you should fill the crankcase with water. The emulsified oil and water mix turns to a foam and will guarantee that you at least get to civilization.

As a wholesale buyer, I made the rounds of every mill I dealt with at least once a year, and with some it was much more often than that. Once when I was in Quesnel, I was about to leave town but realized that I had missed an operator named Brownmiller. I stayed overnight and went around to see him at his plant at about seven the next morning, just as he was getting things under way. He said he wanted to talk and suggested that we go over and sit in his pickup.

It must have been 0° F in the cab but, being a good host, he asked if I would like a drink. Reaching into the glove compartment, he took out a mickey of scotch, unscrewed the cap and offered me the bottle. I took a swig, he took a swig, and we settled into a comfortable discussion.

— * * * —

Maybe 1950 wasn't the old days, but on Saturday nights in Prince George, the RCMP still patrolled the streets in groups of four. When the pubs closed for the night, some patrons went home, some went to regular hotel rooms, and some went to rooms in the hotel's basement that were equipped with cots and called, appropriately, "snake rooms."

I made the 450 mile trip from Prince George to Prince Rupert in the middle of spring breakup, which is the time of year when the frozen gravel roads thaw and turn to long ribbons of mud. The condition of the highway surface varied widely from one district to another, and I am sure that if comparisons were made, one foreman would have been quick to point out the extra difficulties he faced over the smooth sailing in the next fellow's area.

The truth is that even though the railroad went out to the coast early in the century, the roads existed only in short stretches to get from one small town to the next, and it wasn't possible, until the early 1950's, to drive from Prince George to Prince Rupert with any certainty of getting there.

I had the opportunity to peruse some old highway engineers' reports to the minister of public works. They were fascinating to read because they held a mirror to the past.

For instance, back in 1918 and 1919, reports concerning the route that eventually became the Northern Trans Provincial Highway (Highway 16) indicated that the work was confined to maintenance of local roads and trails, and that small sections of new construction were done only to meet the urgent demands of settlers.

In that same period, there was a long report in which the writer seemed to regret the passing of the horse and wagon. He went to great lengths to describe how automobile tires rutted and scattered the gravel road surface, while the iron wagon wheels acted like a roller and packed the surface into a solid mass. History was not on his side.

The loggers got their lumps early when they were accused of finding it more economical to haul logs along public roads rather than build their own skid roads. *"They should not be allowed to haul their heavy loads over main highways, as they have to keep in the centre of the road, sometimes forcing the travelling public into the ditch."*

A reference to Indians in a report hit me with a jolt. It said, *"The great demand for skilled woodsmen and miners at high rates of pay made it impossible to secure men at all. In some cases, Indians had to be employed in order to complete works that were undertaken."* We've come a long way and it's just as well.

In 1931/32 an ominous reference was made to work done on the road from Kitwanga to Hazelton: *"...work was carried out by men on relief on these sections."*

The 1935/36 Chief Engineer's report for the Province contained tables on the number of men in relief camps with the high in 1936 being 6,005 men. The peak in 1933 was 7,783 men and 1934 was close at 7,142 men.

In 1941/42 there was another sign of the times:

"Much effective work was done by settlers working out taxes on roads in the vicinity of their farms."

These reports are a chronicle of the difficulties caused by hard times, bad weather and other natural disasters. I included these comments as a reminder that there was a time when highway building didn't remotely resemble such start-to-finish transformations as the Salmo/Creston, Rogers Pass and Coquihalla Highways that have been completed in such grand style over the last twenty or thirty years.

Getting back to my early trip over Highway 16, it was no surprise that during breakup and the rainy season, there were places where the mud holes were of gigantic

proportions. At some of these there would be an operator with a small cat, and he would hook onto the cars and pull them through. As you cruised through this sea of mud, you could hear those scary grinding, crunching noises going on under your car. True, it flattened the pan a bit, but what seemed to suffer most were little things like grease nipples and license plates.

Once on a hill near Houston, the road was so bad that cars were being towed both ways. We actually had to be pulled down hill.

— * * * —

One of the warm memories of a trip across the north was of staying overnight at a roadhouse in a little place called Topley, which is between Burns Lake and Houston. This establishment was nothing more than a large house with several rooms, but it was used regularly by travelling businessmen. When we assembled for dinner, there was a government forester, a pole buyer (utility poles), a Caterpillar Tractor salesman, and myself. To me, 1951 is hardly yesterday and doesn't rate as the "old days," but that facility made me feel that I had stepped back another fifty years in time.

The woman who ran the place knew the regulars, knew the country, and knew what everyone in the country was doing, so her contribution to the conversation as we ate dinner was both intelligent and informed. Her part seemed to be written for someone like Barbara Stanwyck as the lady boss of the ranch.

The meals were served at a big round table and the food was placed in the centre in serving dishes which were then passed around. How filling and how very comfortable.

One night, on this same trip, I stopped at Endako and stayed in a hotel where they didn't issue door keys and there was no bolt on the inside of the door. Endako was a divisional point on the railroad, and in the middle of the night I wakened to hear much shuffling and talking through the paper-thin walls as the changing train crews

came and went. It was like the Halifax YMCA hostel of 1943. You left your wallet in your pants but put your pants under your pillow.

— * * * —

The lumber industry in the north had its own distinctive way of doing business. Lloyd W. Wise of New York, Broderick Lumber of Pittsburgh, and Edward Hines Lumber of Chicago were well-known everywhere along the north line, and the deals done on a phone call or a handshake would make today's nervous bankers die of fright. Lloyd Wise's office was, and still is, in the Empire State Building. They bought rough green lumber from small mills — scaffold plank, rough industrial clears and 3"and 4"thick commons used for waterfront work and for shoring in sewer laying and other excavating projects.

Orders like these kept many small sawmills in business.

A lot of mills had so much faith in their wholesaler that they would roll cars without an order as soon as they were ready, then they would notify their wholesaler, who would send them an advance on the strength of a bill of lading and a tally. The mill would accept the eventual settlement as the best deal the wholesaler could get. This arrangement of two-way loyalty was very hard for a new kid on the block to crack. So wholesaling was not always a get-rich-quick business, but you could make your own opportunities and occasionally some of them would be lucrative.

Once our company was asked to tender on a very large and varied timber specification, most of which would have been duck soup for any decent sawmill to cut. Because of this, we felt that the lumber could be supplied directly to the customer by a mill at a far cheaper price than we could ever find. However, there was a wrinkle. A large portion of the order was a very difficult specification indeed. As a matter of fact, it was 200 m FBM of fir 6x22 — all 26′ long. The grade was Select Structural, free of heart centres and free of bright sapwood — not an easy thing to find.

On a wild and crazy notion, we put a price on all the items and stipulated that our quote was for all the order or none of it. Our price on the easy stuff was reasonable, but our price on the hard stuff bordered on the outrageous because we didn't have the slightest idea where we would be able to buy it. When the bids were opened, we were the only ones who quoted the tough stuff, so we got the business. It was the most interesting order I ever worked because I could pay over market for the easy stuff to coax a mill to take a portion of the hard stuff in which I had margin to spare. It was the only time I was ever welcomed into J.G. Robson's office at Timberland. John Buchanan usually stonewalled you so that you didn't get to walk in under J.G.'s big stick. (He had a giant natural shillelagh mounted over his door with a brass plate that read, "J.G.'s Big Stick".)

— * * * —

xiiiv) Logs at Hanbury's Mill in False Creek.

xv) Logs and smoke in False Creek.

xvi) Sitka Spruce Lumber / B.C. Forest Products Mill disapearing in a fire in 1959. Area is now apartments, comdominiums and Monk McQueen's.

xvii) Timbers loaded on two flat cars to accommodate long lengths. B.C. fir and cedar on False Creek.

xviii) A typical Interior sawmill. Ganzeveld Brothers, Vernon, B.C., 1951.
Logs sorted by length and diameter. Very advanced thinking.

xix) Ganzeveld Brothers planing mill and yard. The whole outdoors was a dry shed.

xx) "You sell today's production today, and tomorrow's production tomorrow. If you want to gamble, put your money in the stock market." Inventory in shed at Chemainus, 1939.

Chapter 5

The Great North Northwest

One day I found myself running a mill in Terrace.

As a result of discussions I had had with Carl Pohle, he and Jack Webster went into a joint venture on a planing mill, and I was sent to Terrace to manage the manufacturing activities. We bought a planing mill that had been orphaned when fire destroyed Sohen Gill's Yukon Lumber sawmill near Marpole. A fellow named Art Schindel dismantled the machinery and buildings piece-by-piece and we loaded the whole lot into a couple of gondola cars and shipped them off to Terrace. It went back together and operated as though it had never left home.

Carl Pohle was a cherubic little German gent who had worked his way west in 1910, building the Grand Trunk Pacific Railroad, which is now the CNR's north line to Prince Rupert. When the railroad project was done, Carl homesteaded on the Kalum River about five miles out of Terrace. He had a nice country home and, originally, a little sawmill that was powered by a waterwheel. By the time I met him he had built a brand new old-fashioned sawmill in Terrace.

While Carl had some decent equipment and was a canny operator in his way, there were relics of the old days everywhere. Two of these were Walter Williscroft, a millwright who had worked in mills around the north coast most of his life, and Martin Miller, Carl's bookkeeper, who was full of north coast stories that were all new to me.

Prince Rupert, Port Essington, Port Edward, Port Simpson, and Georgetown had existed for years outside the range of my experience, but Miller knew all about them. He knew things that were supported by historical fact, and some legendary things that a lot of people would like to believe. In his version of H.R. MacMillan's early days at the Georgetown Mill, H.R. bought some Davis Rafts of Sitka spruce logs from the federal government because they were no longer needed for war production after the First World War. These were supposedly bought at bargain prices because of teredo worm infestation that just may not have gone as deep or into as many logs as H.R.'s bid suggested. On the other hand, I have heard that they were full of pink stain which is an incipient decay and that H.R. nearly lost his shirt on them. So take your pick.

A Davis Raft consisted of logs piled and formed into a huge bundled raft containing as much as three million feet of logs. These were tightly bound together with heavy cables and chains to allow safe transport through the heavy seas from the Queen Charlottes to the Lower Mainland mills. In due course, Davis Rafts were replaced by self-powered, self-dumping log barges.

— * * * —

At the Terrace mill I had a thorough exposure to fundamental religion. The crew was divided into significant groups of Pentecostal Assembly, Dutch Reformed, Jehovah's Witness, and Seventh Day Adventist adherents, with a few old line denominations thrown in. It was very hard to get a quorum on the weekend. The Dutch Reformed kept Sunday rigidly. The Seventh Day Adventists kept Saturday; they also made things a little awkward at sundown every Friday afternoon in the winter when they had to be replaced for an hour or so. Nothing would stop them from leaving work at the time the church officially designated as sundown. There were balancing advantages, though. Seventh Day Adventist millwrights would work Sunday if there were jobs left unfinished on Saturday night.

One of the little Sunday chores was to grease the bearings on the spool on the end of the waste conveyor that entered the side of the burner and hung out over the fire. Obviously, this had to be done when the fire had cooled.

One Sunday, when Miles Bell was in to do this chore, he drove his car into the yard and parked, leaving his baby daughter asleep in the back seat in a bassinette. He had parked his car in the shade beside a lumber pile that was piled three loads high. For whatever reason, the lumber pile fell forward and landed on top of the car. The roof was flattened, the springs were compressed so that the wheels were right up in the fenders, but the baby was sound asleep in her blankets, quite unharmed.

If there is a plan, it is obscure. This same child, as a teenager, was riding in a car with some other kids when the driver overtook another car on dusty, gravelled Highway 16. They pulled out into a cloud of dust and passed right into an oncoming logging truck. The kids were all killed.

— * * * —

While I was in Terrace, I took my first lesson in small town diplomacy (I took a few). This is an essential commodity without which your life in a small town will be calamitous, miserable, or short. In the Terrace of that time, there were still many pioneers and sons of pioneers living in the town and they all had well-known and colourful backgrounds. Their activities were the result of many years of "one thing leading to another" and anyone who didn't know their history automatically labelled himself as a carpetbagger.

One of these descendants of the old guard was a logger. He approached me at the Pohle mill to see if I was interested in some spruce he was going to log on one of the islands in the Skeena. We agreed to a trial truckload which came in round and sound and straight, and on the strength of this I decided to take the production of that particular show.

As it turned out, the logs were never quite that round or sound or straight again; in fact some of them you could have called "crappy." Some had knots six inches in diameter from limbs that must have been growing out in the open like the trees in a park. Grumbling always brought an optimistic report: "Don't worry, the next few loads will look a lot better than this."

When this was all behind us, I happened to tell my woes to a fellow who ran the hardware store. The way I put it was, plainly, that I thought I had been skinned. The storekeeper said, "Oh, I find that hard to believe. Did you know he is my brother-in-law? I'm sure he would never do a thing like that."

Regarding that principle of "Let's-try-a-truckload-and-see-if-the-stuff-is-any-good," there would come a time when I tested the theory practically every day, buying lumber as raw material for a remanufacturing plant.

The idea was eminently sound, but the results were

often the same as those above. The response to my anguished queries on the second and third loads usually had something to do with my being over 21.

— * * * —

At Pohle's I had a brush with basic civil law, although nothing came of it. We had a steam plant and had a well that was used as a reservoir of make up water for the boilers. This well was kept full with water from a spring that flowed out of a side hill, one of the terraces for which Terrace was named.

During a dry summer, this spring disappeared without warning and so we had to do some prospecting for a new supply. It wasn't far away. We dug a small tunnel where our spring had been, and suddenly hit a gravel seam out of which beautiful fresh water gushed.

About a week later, a farmer from about half a mile away came in and said, "Say, have you guys had any trouble getting water? I've run dry." We had violated the downstream users' water rights.

— * * * —

Pohle's mill was another that had a steam engine still doing its duty. An old gent named Caruso was the steam engineer and he kept the engine room looking like a museum model. There was a 10 foot flywheel with the spokes painted red and the thing just whirred along. Since there wasn't much excitement in the engine room, the engineer spent a lot of time shining brass and sweeping the floors. The fireman in the boiler house at least got to sweat.

Carl Pohle was a short pudgy man with a perpetual smile on his face. He was bald and had a habit of rubbing his hand over his pate, sometimes removing his hat to do it. Being somewhat in the same condition myself, I can relate to his impulse. Not only is it something to do with your hands, but it is undoubtedly the remnant of some noble gesture of the past, possibly relief at getting your chain mail off or upon removing your armoured helmet.

But Carl had the habit of giggling. It was a nervous mannerism so he even giggled in the face of peril and disaster. He giggled when I left a forklift parked on a patch of ice at the brow log where the logs were unloaded from trucks to tumble down into the log pond. While we all watched, the forklift slowly and gently slid toward the edge and then tumbled dramatically down to the bottom, breaking off one of the rear wheels as it went.

Carl had been in Terrace so long that he had bits and pieces of private timber all over the place. One of these was close to town and you'd have thought it would be the first to be logged. The trees were large and straight and on the whole quite impressive. But they were also overmature, covered with conchs (conks) and not worth a whole lot as sawlogs. In the absence of normal usefulness, they performed a different but nonetheless vital function. Carl left them standing to impress his banker whenever the question of collateral came up.

Being German, Carl was a good contact for new German immigrants. Some of these had been in Canadian prisoner-of-war camps and one had been in the submarine service. This fellow and I became good friends and when I found out about his service, we spent a few minutes speculating on whether he might have been in the sub that torpedoed a frigate I was in off the Azores in 1944. It was probably something we were both better off not knowing.

— * * * —

Bill McRae was a young man who had married the boss's daughter, Helene Adams. Mr. Adams had a mill at Usk, where he logged on Indian Reserve property and in turn employed entirely native people in the bush and in the mill.

Bill had been a powder man in hard rock mines and tried to conceal this when he joined the army in World War II. Somehow the word got out, and Bill spent most of his war years doing hair-raising demolition work under the enemy's noses.

Bill and I and Gerry and Helene picnicked one early summer afternoon on the Copper River. The effect of hot sun on the snow on exposed hillsides is not to be underestimated. We carefully placed our beer among some rocks in the river and went for a hike. When we returned in mid-afternoon, the water in the river was two feet higher than it had been and the beer had floated merrily down the stream.

Mr. Adams' mill was right on the Skeena River.

When the salmon are running up to spawn, there are hundreds of cutthroat trout in the river, too, going up to gorge themselves on salmon eggs. We stood at the edge of the river baiting barbless hooks, nothing more than bent pins, with salmon eggs, and we jigged trout, flinging them out onto the bank.

Terrace and the whole north country is fly country. I have seen Indian loggers and pole skidders give up and come out of the bush because the no-see-ums were so bad. Your first exposure to black flies is a bit alarming too, since they desensitize your skin so that you won't know they are biting out a chunk of you. Your first hint that you have been bitten is when you see a little pool of blood on your arm — probably some that dribbled out of the corners of their greedy little mouths.

— * * * —

Gerry and I made several trips to Prince Rupert and established ourselves as the luckiest people alive: every time we went there the sun was shining. Not many people can say that.

On an earlier trip. when Art Schindel and I went up to Terrace to case the prospects of the planer mill construction, we landed in Prince Rupert in pouring rain. We enquired about a rental car and found that the Undertaker had the only rental cars in town. Presumably he found it convenient to have another use for them when days were slow at the parlour.

We thought the car he supplied smelled a bit musty when we got into it, and once we headed out on our 97-mile trip to Terrace we found that the heater and the defrost did not work. It thundered down rain for the whole trip, and while one drove, the other wiped the steam off the inside of the windshield. We found out later why everything was so dank and smelly: shortly before, someone had run the car off the highway into the Skeena River where it was completely submerged until it and its expired occupants were found and fished out.

— * * * —

At Remo, which was only a name on the railroad a few miles west of Terrace and at the mouth of the Zymagoticz River (locally called the Zymacord, and small wonder), there was a perfectly preserved old sawmill. It had been built for Sir Henry Pellatt, the Eastern Canadian industrialist who also founded the Canadian Fishing Company.

This mill was built in a large back eddy on the Skeena and the scheme was to bring logs down the river and hold them in this still area until the mill used them. The river had other ideas. From time to time, the Zymacord would decide it needed a change down at its mouth and would take up some long-disused route for its final entry into the Skeena. This made the locating of a permanent mill site a bit of a gamble, and as it turned out, Sir Henry's mill wound up on a gravel bar half a mile from the closest water.

The mill should have been preserved as a museum, though I'm told that it was demolished a few years ago. It was steam operated, with a huge carriage that could cut an eight foot diameter log and sixty foot timbers, although their main product was lumber for fish boxes. The carriage deck looked as big as a ballroom.

The strangest thing about the whole setup was a planer that was anchored and grouted into a sloping rock outcropping underneath the sawmill. The lumber fed uphill, through the planer and up a belt through the mill floor to the carloading deck. I wouldn't tell you such a yarn if I hadn't seen it myself.

Sir Henry's fascination with rocks may have carried over into his castle — Casa Loma in Toronto, the famous eccentric rock pile that fifty years ago was turned into a resort and was the summer home of Glen Gray and his Casa Loma Orchestra.

All of the equipment in the mill was built by Waterous of Hamilton, and it was a toy to bring tears to a sawmill man's eye.

When I was there, this mill was owned by a man named Claire Giggey, who kept it cleaned up and who even painted the smokestack from time to time. And all of this was surrounded by trees, out of sight of all the people in the world except for perhaps the half dozen residents of Remo.

— * * * —

Claire Giggey's other activity was the logging of the giant Black Cottonwoods that grew along the Skeena and on the islands in the river. Anyone who has had anything to do with cottonwood will know that it is too heavy with its own water to float. Giggey loaded these logs on rail cars and took them to Prince Rupert. There he made a cradle of cedar logs woven with wire cables onto which he dumped the cottonwood logs. The rafts were then towed to Vancouver where they were peeled into veneer at Western Plywood, at the foot of Fraser Street.

In 1951 Columbia Cellulose, which was the forerunner of BCRIC and now Repap, logged in the Terrace area and hauled logs by trainload to the pulp mill at Watson Island. In places, Highway 16 and the railroad shared a narrow flat spot beside the river, cut out of the rock at the bottom of the steep mountainside. Driving along beside a lurching log train was something to bring your heart to your mouth.

Something that never changes is the intense quiet of a wide flowing river. There are places on the highway between Terrace and Prince Rupert where the Skeena is a quarter of a mile wide and where you can stop your car and listen to the silence. The only moving thing might be a soaring eagle and no matter how you strain to listen, there is not a sound to be heard. The river moves, the surface boils gently to show some character, but the sound is turned off.

Once one's chaotic duel with the train was over, a visit with this quiet river was a pleasant relief.

— * * * —

Terrace is not only east of Prince Rupert, but north as well, and in mid-summer the days are very long. We were of the right age and attitude to party when we were up there and we really did go out in daylight and come home in daylight.

When we were new in town, we were asked to a party at a neighbour's home and unwittingly arrived before we were expected. We caught the hostess in her slip, which is pretty standard, but the host's preoccupation was unusual. He had their goldfish in a bucket on the kitchen

counter and was scouring the fish bowl to get it ready for punch bowl duty.

We were in Terrace just when they were building the railroad down into Kitimat. With the long hours of summer daylight, the construction company worked practically around the clock. The drone of the diesel electric

Letourneau belly loaders scraping away at the right-of-way was never out of our ears. There was a cut they were making in the blue clay out near the airport and the material was so gluey and heavy they had to push the scrapers with a TD 24 bulldozer to help them pick up a load.

— * * * —

We played poker on a regular basis with friends. One Saturday afternoon a well-known wholesaler happened in. We fed him supper but explained that we had a regular Saturday night activity. He said poker was his favourite game and he would like very much to join us. In addition to being unlucky, he turned out to be a bluffer and a hopeless optimist. We didn't play wild games so it was fairly easy after awhile to figure out his style. He lost quite a bit of money, but he kept on smiling. I'm afraid he did business the same way, so his career hit a few rocks, purely from being too impatient to get rich slowly.

One of our poker-playing friends was Neil McKerracher, who was the town barber. His shop was the clearinghouse for all the latest gossip and wisecracks. Neil had a real chuckle the day an ancient Chinese gent came in and passed judgement on one of the local businessmen. "Ah, that Gerry," he said, "mind like a child."

— * * * —

A prominent local lawyer had a remarkable thirst and an athletic avocation. In his drinking days this fellow was an unusual baseball umpire. He came to the game equipped with a bottle of coke well mixed with rum, which he kept in his hip pocket. The farther along the game went, the more flamboyant his calls and his gestures became. This was all part of an act he had worked up, and the more he cavorted, the more the crowd egged him on. He was better entertainment than the ball game.

— * * * —

Long cedar poles are fragile and sometimes they were pre-logged ahead of the sawlogs to avoid breakage. The loggers used horses to skid these poles, and in the winter these horses were turned loose and allowed to graze at will. Yards and boulevards became open range, and often they would get into a yard and stand in the lea of the house to keep warm. After awhile you knew what it was when you heard a horse stamping his hooves for whatever reason it is that they do that.

One night my wife woke me and, with obvious concern in her voice, whispered, "Wake up; I think there's someone in the house." I did my brave man act: got up, put on my dressing gown and, with legs of jelly and a racing heart, ventured down the stairs. I called up to

Gerry, "Come on down. You've got to see this." A horse had nudged the front door open and was part way up the steps so that his head and neck were in the house.

Cattle had the same freedom, and one day Gerry was incensed to find that four of them were in our front yard. Being a city girl, she grabbed the first thing she could lay her hands on — the fly swatter — and tried to haze them out of the yard. This only caused them to scatter in four different directions. Lewis Smith, the kid next

door, yelled over, "Gee, Mrs. Gatensbury, don't you know they have their own fly swatters?"

— * * * —

xxi) Herman Buschman, Dick Higgs, and the Author.
Pohle Lumber Co., Terrace, B.C., 1952.

xxii) Sitka spruce logs at Pohle Lumber. Gin Pole, Cherry Picker, and various other contraptions yarded logs to the bottom of the log haul. Not high tech.

xxiii) Pohle Lumber Company, Terrace, B.C. Refinements to come.

xxiv) Timbers cut at Pohle Lumber, Terrace, B.C., 1953.

Chapter 6

The Kootenay Loop

In 1953 Gerry and I decided to leave the north. I was caught in the middle of a growing disenchantment between Jack Webster and the Pohle people who thought that Jack's orders were not as suitable for the mill as the ones Wise and the old guard had provided them with in the past. My loyalties had to remain with Jack, but as I said, I was caught in the middle. I had answered an ad for a sales manager's job at Kootenay Forest Products in Nelson, and when we got down to the Coast, there was a message from Nelson to call Vern Vance.

We headed off to Nelson to size up the place and for me to have an interview. Vern's news was, "Gee, we hired a sales manager, but we need a superintendent, too. Would you be interested in that?" I said, "I'm not doing anything. Why not." The new sales manager was Ray Wallace, and we spent four fairly uninhibited years together.

KFP was a relatively new mill. In 1953 it had only been operating two years and had gone through some severe growing pains. I was the fourth superintendent in two years, and whatever else I may have added, I did bring some stability to the plant. When I arrived, my predecessor had not yet left town. Gerry met his wife, who gave her an earful on what a terrible outfit it was to work for. This turned out to be not true, but it was an unusual way to start.

Most West Kootenay mills saw at least eight species, namely fir, larch, hemlock, white pine, cedar, white fir, Englemann spruce and, to a small extent, Ponderosa pine. This makes the business interesting. It is a salesman's dream or nightmare, depending on how much he likes a challenge.

KFP was privately owned by the Leuthold family from Spokane, Washington, until it was sold to the Eddy Match Company, who sold it to NDP Provincial Government/ BCRIC/Westar, who shut it down in 1978 and demolished it in one of the dumbest moves this industry has ever seen.

Eddy Match, through a series of mergers between various British owners, became a subsidiary of Wilkinson Steel. This company had no interest in a Canadian timber operation, and put KFP up for sale. Crestbrook Timber, whose timber holdings on Kootenay Lake were interspersed with those of KFP, offered to buy the company as a going concern, but, because of Crestbrook's Japanese partners, the Barrett government would not allow the sale to go through, and bought it themselves. The eventual result of this was the shutdown and the loss of approximately 400 direct jobs plus the service jobs they supported, and this was something Nelson could ill afford.

When I was there, the Leutholds still owned it and they were good operators. They also owned Deer Park Pine in Deer Park, Washington, which was a Ponderosa pine plant with a sawmill and complete manufacturing plant, where they made everything from Colonial furniture parts to barrel heads to toilet seats to Wolmanized window frames to toy blocks and even blocks for shoe heels for women's shoes, with what was left going to Presto logs.

But getting back to KFP, the management arrangement there defied the textbooks because Bill Shukin and Vern Vance appeared to be joint managers. Vern was a smooth salesman who was just as at home in the bush as he was in an office. He was also a raconteur of some note.

Bill Shukin was the general manager and designated money-worrier, and came from a family with roots in the earliest Grand Forks Doukhobor community. Between them they ran a good company that made the most of a log mix and quality that was a tough challenge.

— * * * —

When the mill was new, before it had even sawn a log, the Spokane owners came up to have a look, and one of the Deer Park sawyers had asked to come along. He wanted to see what a new sawmill looked like.

When they got to the mill, the sawyer asked if he could pump the rig — move the empty carriage up and down the track just to get the feel of it. The carriage drive was a gunshot feed, which is a long steam cylinder with the piston rod connected directly to the carriage. In its day, this was the fastest and most responsive carriage drive there was.

In a sawmill, it is obvious that the logs will either come down from a log deck on the sawyer's left or right hand side, and the rest of the mill will be on the side away from the log deck. The Deer Park sawyer was accustomed to sawing on a rig of the opposite hand, and when he sat down in the box and moved the stick as he would at home, the rig seemed to him to be going the wrong way. Seeing this, he pulled the stick even harder in an effort to get the carriage to stop. As a result he shot the carriage down the track, tore out the bumper, and sent the carriage halfway out the end of the sawmill building. At the same time, he pulled the head off the gunshot cylinder and filled the mill with a roaring blast of steam. After that opening, anything else that happened at KFP would have to seem tame.

On the outside of the building there was a stack that vented the exhaust from the gunshot. Someone had closed in the top of this stack and cut openings in the column in the shape of eyes, ears, nose, and mouth. This vented the steam and was a conversation piece. It probably vented as much wisdom as some of the experts that were to follow.

— * * * —

Vern was a storyteller and he was also a social entertainer without compare. Vern's wife, Teddy, was always prepared for Vern to call in the middle of the afternoon and announce that he had some surprise visitors who he would like to bring home to dinner. Teddy had an alter ego, Mrs. Jordan, whom she kept on a retainer. There was never any panic. She just called Mrs. Jordan, who saw that everything got done right and on time.

Mrs. Jordan raised the Vances' six children on a proxy basis, cleaned house, cooked and sewed, but never lived in. Teddy, for her part, composed herself and was a perfect hostess.

Gerry and my girls, when faced with a domestic dilemma of any kind, still say, "Why don't we just call

Mrs. Jordan?"

— * * * —

Some of the social gatherings at Vern's involved his lumber-buying cronies, who were not known for their sobriety or any other restraints on their boyish enthusiasm. Vern's house was on a hill, as are all Nelson houses. As a matter of fact, the climb from street level to Vern's house was twenty steps followed by a sloping walk.

One cold winter night, after a bunch of us had lingered long over dinner in town, we came back to Vern's place for the rest of the evening. At the top of the steps, Monty Jewett spied Bobby Vance's Red Racer. He burbled, "Gee, I haven't done that in years," he took the sleigh, took a run, flopped down, and slid across the lawn and down the steps head over heels to the street.

Most of us were in the house when Monty arrived, horizontal, with a man under each arm and each leg, bruised and scuffed but with a silly grin on his face. He did not say, "If I had it all to do again, I'd still do it." He was stewed but not stupid.

One night we had a long dinner at the Round-Up Room, which was a better restaurant that its name would suggest. It was one of those nights when Sugar Ray Robinson was fighting Kid Gavilan.

There were some pretty reckless bets being placed, but the goofiest part came when someone observed that we were betting on a fight that had been over for an hour. We had lingered so long at the trough that we had forgotten to tune in in time to hear it.

Vern told about a time when he was buying Lodgepole pine pit props. These were small rounds, eight feet or so long, that were bought to replenish the needs of the British coal mines after World War II. These props were being cut on what was then a very lonely back road that went from Rock Creek through Beaverdell, Carmi, and McCullough, to come out near Rutland.

Vern and a working companion had to stay overnight in a small hotel in Beaverdell. When they retired, the other fellow removed his dentures and put them in a pitcher of water. Mischief was afoot. The weather was cold and the pitcher was near an open window. They set their alarm wrong and left themselves short of time to rise and drive to Rock Creek to catch an early morning Kettle Valley train. In their rush, Vern's friend was horrified to find that his teeth were frozen solid in the water pitcher. They made it, but he did the only thing he could do under the cir-

cumstances: he boarded the train with the pitcher in his hand.

— * * * —

The shipper at KFP was a man named Warren Hardy, known to all around as "Waddy." Waddy was an Anglophone from the Gaspe Peninsula so he had a bag of linguistic treats that were both rustic and French flavoured. Waddy also had a keen but wonderfully dry sense of humour.

In the down home Gaspe vernacular, someone had observed, "Them's good molasses." To Waddy, that was an all-purpose commendation.

Waddy told of a logging company employee who was

sent to assess the quality of a certain block of timber, and when he returned, his answer to the obvious question was, "She look pretty good from the road."

Waddy had a cousin who lived at home with his family. One Saturday as he was leaving to make his regular weekend trip into the nearby small town, he tossed an aside to his mother, "Put another piller on my bed tonight, Ma; I'm gittin' married."

Waddy lived a double life. Well, maybe not double, but it came in two parts. He had been around and had survived a full stint of army service in Europe. His stories of life in the Gaspe pulp logging camps were grim humour at best. He had lived his life in a rough, male-oriented world, but he married a gentle religious lady who lived for the spiritual comfort of the Pentecostal Church.

Waddy would occasionally get together with the boys after work, and if the evening wore on and the oil was flowing freely, he would be afraid to go home. We might be all ready to go to bed when there would be a light tap at the door and there would be Waddy, looking for a cup of coffee and a chance to regroup. These nights were cheap at twice the price because he would be wound up and the stories would flow.

He told about the depression days in the Gaspe when he, his brothers, and the sons of his many relatives would go from farm to farm, pooling their labour. His uncle John had no kids and the boys hated to go there because John's wife was a lousy cook. She would send them out into the fields with a lunch that consisted of sour cranberry pies made from the berries that grew wild in the local bogs. What is more, they were piled on top of one another so that they all ran together.

One day the boys dropped their pretense and declined to eat. When they got back to the house at the end of the day, Uncle John announced, "The Hardy boys don't like your goddammed cranberry pies!"

Waddy was the only one of his family living in the West.

When he went home to the Gaspe it usually meant a family gathering at which he would be the centre of attention. Once when he was speaking enthusiastically about 10,000 foot high mountains, thousand mile automobile trips and lakes a mile wide and a hundred miles long, he noticed that one of his uncles remained strangely quiet and unsmiling. Finally at a break in the conversation the uncle spoke up: "Your father was a liar and you're a goddam liar too."

One of Waddy's men was given the task of canvassing the crew for some vital statistical information. He prepared his own form and made the rounds with his pencil and clipboard. His headings were:

LEGIBLE : ELIGIBLE

In Jack's unusual classification system, the Legibles were "in" and the Eligibles were "out".

Waddy wasn't a hardnose but he did expect things to be done right.

At KFP we shipped a lot of what were called "lot loaded" cars. These were rail cars that contained lumber for five or six different customers, often in different cities. This meant, for one thing, that after the top lot was removed at the first stopoff, the remaining material had to be stowed in such a way that it could travel without damage to the next unloading point.

But getting the picture straight in the first place was most important. The order would clearly state: "Lot One. Load first on the bottom, to come out last at Muncie, Indiana. Strip off and mark: E.A. Allen, Muncie, below."

The next lot was then loaded on top of the first and then the next, until everything was in, with the lumber for the first stop on top. The instructions were particular and nothing was left to chance.

One day, tallyman Johnny Kabatoff came in from the cardeck with a big sigh and said, "Well I really effed that one up. I loaded it upside down."

Johnny was a good man and Waddy knew that you could strike out once in a while and still have a good batting average so Johnny's life was spared.

— * * * —

Life is not all jazz and joy. One day, right at noon, when I was still busy with something in the mill, Waddy ran in pale and agitated. I immediately asked, "What in the world is the matter?" He answered, "The trimmerman just jumped in the hog." A hog is a heavy machine with rotating hammers, used to pulverize wood waste material. In our case this material was blown straight to the fuel bunker and eventually into the fires that heated the boilers. That was where this man had gone after removing his jacket, his spectacles and his wallet, and leaving a note. I will spare you the grisly details but it made for a different day.

— * * * —

KFP was a happy mill for the most part. As supervisors we certainly were happy. For whatever ill-advised reason, we still had an all-male stag party once a year, and these events were roundly hated by our wives. For our part, we did what we did best: we ate, drank and gambled on our night out.

One year we formed a quartette, and we got to be pretty good. We were Ev Rolph, Laurie Lefeaux, Russ Hawes and myself. Among other things in our repertoire, we dug out that old temperance groaner, "Father Dear Father" (come home to us now/the clock in the steeple strikes three). It was probably a bad choice to sing at the stag because most of the guys in the audience were living the words as we sang. I remember that it got awfully quiet.

I said we were a happy company. Well, let me modify that. There was a pervasive feeling among the mill people that the loggers had a license to print their own money. The loggers, in turn, took a defensive stance that anticipated a momentary attack on their access to the goodies. Of course, as mill men we probably didn't know what we were talking about, but it looked like they were getting away with murder, and that was enough for us.

One year for entertainment at the stag party I got the idea (misbegot it) to rewrite Robert Benchley's "Treasurer's Report" in terms that would be a little more familiar. Bill Shukin, the manager, volunteered to be Benchley and armed with a chart and a pointer, he proceeded to present a wacky report, item by item, until he got down to Benchley's enormous "miscellaneous" entry which, in our version, was the logging manager's expense account. This little jest landed like a bomb and several of the loggers stormed out with black scowls on their faces and unkind words on their lips.

Having set the mood, things got to a point where,

when one of the plant foremen arrived wearing a red bow

tie, something that the logging superintendent thought was his own personal trademark, the log super thought he was being mocked and he, too, left in a huff. Yes, there was that little matter of rivalry between the mill and the loggers, and it probably led to the eventual conversion to a subdued mixed company function. Or maybe it was because someone tried to throw a piano out of a second storey window, though that was after my time.

— * * * —

Jim Harvie, an old friend from Vernon, followed me up to Terrace and did my old job there for a year or so. One day I got a call from him with the message that he had had enough, and the question "Do you know if there is anything available around the industry that I could handle?" I offered him a mill foreman's job and he surprised me by taking it.

Jim arrived in Nelson on a Sunday and when I met him at the office, I said, "The first thing we have to do on Monday morning is to get the master mechanic out of jail." (He was only a felon to the point of sleeping one off in his car with the keys still in the ignition, but he did it with some regularity and the police had his number.) Being fresh from the Terrace Bible belt, Jim's reaction was predictable. "That's more like it," he chirped. "Jesus, I haven't even been able to bum a chew of snoose for over a year."

— * * * —

Jim was raised at Willow River and Giscombe, east of Prince George. I could listen to Jim talk all day about horse logging and log sleighs and ice bridges and the details of practical backwoods survival. He tells about shaving chips of moosemeat off a frozen carcasse and frying them quickly in a very hot pan — northern junk food. Northern depression jam was made of turnips, sugar, red vegetable dye and some alfalfa seed to make it look like strawberry.

Dumb, to Jim, was not knowing the difference between a bull whip and a wagon wheel.

When Jim and his friends were old enough to want to go into Prince George on Saturday nights, his mother admonished them, "Remember, if you lie down with dogs, you get up with fleas."

Jim was a fisherman. He knew the right lure and the right fly for every hole on Kootenay Lake. The fish knew that he knew and stayed out of his way.

We visited them in later years, and Jim and I made

a furious assault on the best spots, applying the best of his accumulated lore, but with no results.

One evening we visited with other friends and when we came home to our beds at the Harvies' house, there was a note, in Mary's writing, pinned to my pillow. It said, "Steve, go look in the fridge." There, lying across the front

of a shelf, was a rainbow with its nose up against one side and its tail curled up the other.

Mary's explanation was that she didn't know much about fishing. "I just drag an old wet fly around behind the boat and they seem to bite."

— * * * —

There are some rights that are sacred and there is no way they must ever be tampered with. One of these is a man's seat in the lunchroom. At KFP, there was no cafeteria. There were rows of seats that had one wide arm instead of a lunchroom table because it was strictly a lunch box situation.

There was a guy in the shipping crew who got under most people's skins because he was a bit of a dude. He was a big guy but he wore his hair long, well before long hair came into style. He smelled of perfumed talc, played the guitar, and sang western music, and talked about his talent incessantly. A bad combination.

One day, George Madelung came into the lunchroom to find the dude sitting in his seat. The following ensued:

George: "You're in my seat."

Dude: "I don't see your name on it."

George: "Well, it's mine and I'll give you three to get out of it. One. Two. Three."

With this the dude got up, smacked George on the jaw and knocked him onto his backside on the floor, and then sat down again in the same seat.

George came into my office and asked me what I was going to do about it. As far as I could see we were dealing with:

1. An unwritten law;
2. A challenge;
3. An asssault.

One was unsupportable; two was foolhardy; three was either self-defence or jungle rules, but it was a judgement call. I don't remember how I smoothed things over but I'm sure it was plenty platitudinous.

— * * * —

One of the first problems I faced at KFP concerned sawfiling. They were making poor lumber and the filer had convinced himself, and almost had the manager convinced, that the band mill itself was to blame and there was a need for a major repair that would shut the mill down. I looked at what we had, talked to Phil Quelch from Simonds Saws and a few other connections in the know, and then told the filer, "You may be right, but the company deserves to let someone else try."

We hired one of the Portlance brothers. Leo, Bud and Eddie Portlance had the reputation of being the "royal family" of filers. Ed came in and our problems were over (almost).

Ed was one of those odd birds who spent every weekend roaming over what today we delicately call land fills. He scrounged copper wire and scrap metal and four-legged stools with only three legs working, and other such collectors' items that he repaired in the filing room, and it soon appeared that his whole attention was not on our saws.

As a result, in six months we were back where we started. We pulled off a devious scheme to have brother Leo come up to visit Ed and just incidentally sneak a look at Ed's handiwork while he was at it. Leo's assessment: "There's nothing wrong here that a pair of glasses wouldn't fix."

Ed Portlance was eventually followed by Cece Alexander, who was one of the wackier characters in the business. At KFP he filed for three band rigs, a solid-tooth four-saw edger, a 24′ random length trimmer, and a log deck chain saw and a slasher. He had one fitter in the filing room, which is about as few as you can get.

Cece did bench and weld his saws, but that was all he did. He ran the tail off his fitter whom he had trained to do every other job there was in the filing room. The fitter could have cheerfully killed Cece because the latter had set himself up with a soft-cushioned rocking chair, an end table, an electric coffee pot, and a rack for magazines and other reading material.

One day I couldn't hold back any longer and I said, "Come on Cece, that looks terrible. What if someone walked in and saw you set up like this?" He said, "What's

the matter; aren't the saws running right?" I said, "Yes, they're running O.K." He said, "Well when they aren't, I won't be sitting here."

We got a year end bonus at KFP, and shortly after Cece got his first one, he made a trip to the office with a plan whereby the next year's bonus would be divided by twelve and paid out at so much a month, this before anyone knew if there would even be one. Vern's predictable reaction was, "Geddouddahere."

Sawfilers do not talk a lot about their work so in order to be able to sort out the facts, I eventually became very well acquainted with the technicalities of filing.

Much later, when a group of us with our wives were being toured through the magnificent new B.C.F.P. cedar mill at Hammond, we were in the filing room and I was explaining to Gerry how the tips of the bandsaw teeth were "squashed" flat to make them wider than the rest of the saw blade, this being the secret of a freely-cutting band saw.

Doug Clitheroe was within earshot and he said, "Swage, Steve, swage. You swage the teeth to widen the tip." I said, "It's 'swage' to you and me, Doug, but it's 'squash' to Gerry if I'm going to make myself clear."

— * * * —

There is no question that the company's timber existed in patches here and there all over the Kootenay Lake watershed. There was a mysterious private timber stand at a place called Tye, that got a lot of publicity and was sold frequently from one outfit to another without ever being logged. It was inevitable that Bill and Vern would be conned into making an offer on this timber. They made an earnest money deposit and then made the long trek in to see it. As they put it later, "We trudged up, but we ran all the way down to get our money back." The timber was spread over a vast area of rocky mountainside but it looked pretty good on paper in Seattle, which was where the owner was promoting it.

When I left KFP a few years later, the walls of the home where they gave me a party were decorated with posters. One of these was a depiction of Tye. It was a steep mountainside with one tree, three pyramids, the sphinx, a steer's skull with horns and eye sockets, and two camels.

All of Kootenay Forest Products' logging shows seemed to be steep and rocky.

Cedar poles were a premium item. The only time we ever got good small cedar sawlogs was when the hairpin turns in the logging roads were so tight that they couldn't get loads of 50′ to 90′ poles around the corners. They would then have to buck the trees into short logs for delivery to the mill.

— * * * —

Not all the entertaining was done at Vern Vance's place. With the proper amount of notice, a couple named Gehrig would serve an elegant gourmet dinner in the dining room of their home, which was perched high on the top of the mountain behind Nelson. They consulted you on the menu and then went out and bought precisely what they needed for the occasion.

Without fail, Gehrig would appear at the office the following morning, as soon as he had tallied up the bill.

— * * * —

KFP was on the lake and the logs were boomed in big round bag booms and towed to the mill by the company's tug boats. At one of the camps, the logs were put in the Kootenay River in Canada, then towed in a section of the river that looped down through the States before coming back into the lake in Canada at Creston. It wasn't as easy as it sounds because at one point they had to pass under a bridge and this required that the bags be narrowed up with cable swifters, which was a lot of work.

When the logs came up the jack ladder, there was a pressure washing apparatus which cleaned off dirt and rocks left on the logs from skidding them in the bush. It also knocked off bits of bark and, with that, numerous types of bugs and larva, all of which ran back into the lake. The fish came for breakfast, lunch, dinner and

snacks. The boom men had set lines that had an eight or ten inch section of elastic band spliced in to give the line a soft rod tip action, and they took their chances on whatever punishment the Fish and Game officials would have meted out if they had been caught. Have you ever seen a bucketfull of Kokanee?

— * * * —

Some things never change. Gerry shared a seat on the bus with one of the setters who was off to the theatre for the evening. What he didn't know was that I had been late getting home because of juggling the crew to accommodate his absence, and Gerry was well aware of this. More than once the union has patiently explained to me that a man may be too sick to work but well enough to do something light like going to a show.

I always felt better about the guy who called in and said, "Gee, we were celebrating my Aunt Millie's birthday last night and I'm so hung over I can't make it in."

— * * * —

I haven't bothered to mention Doukhobor days and Doukhobor ways because when you lived in the Kootenays you had a much more balanced view of their situation. As one of their members wisely noted, "When a bomb goes off in Nelson, it makes a much louder noise in the offices of the Vancouver Sun than it does here." We did make occasional trips to the Slocan Valley and other places where these people lived — Goose Creek, Winlaw, Vallican, Slocan Park, Perry Siding, Thrums, Shoreacres, Brilliant, and Castlegar. We had good friends among them and ate several of those prodigious meals of borscht, chicken, corn-on-the-cob, vegetables and boiled or fried pyrogys. The support stuffs were pitchers of melted butter, sour cream, and heavy sweet cream to go on your baked apple pyrogy (yes, baked).

— * * * —

There were squatters who had established some shacks at the edge of the lake above the mill property. We had decided to put a cleanup man on a graveyard shift in the mill to double as a fire picket. The clanking and squeaking of various conveyors kept the squatters awake and it wasn't long before a delegation came down to the office to complain about the noise. Since they were technically on mill property, they didn't get much of a hearing.

We did have guard dogs, though. The Kettle Valley branch of the CPR went right through the plant, splitting the sawmill and dry yard off from the kilns and the planer mill. This allowed our squatter friends and other transients to walk up the track and right through the mill property. The dogs were chained at each end of this throughway and I'm sure this discouraged many people, but I lived in fear of some trusting child being attacked. The dogs were trained by the RCMP and if you have ever seen what a German Shepherd will do to a trainer who is wearing padded clothing, you would think twice about intruding.

— * * * —

Along the arm of Kootenay Lake and much of the main lake, the hillsides are covered with a lush green carpet of young forest. Vern Vance said that these areas were never logged, but instead had been burned off in forest fires in the mid 1920's. Vern had worked on the CPR stern-wheeler lake boats when he was a teenager and he said that at times the smoke was so thick that you could not see either shore of the lake from the boat.

— * * * —

When in Rome....

When we were in Nelson, the bridge to the North Shore had not been built. There was a ferry that pulled itself back and forth on cables, and, yes, it was possible to miss the last ferry at night.

Even in that sleepy little town there was a ferry lineup at the peak periods of the morning and the afternoon, and an absolutely unbreakable rule was that you never, never cut in at the front of the line.

Leo Atwell, who had been a pretty good hockey player in the old Western International League, had retired from hockey and was working for a finance company in Nelson. One evening the unspeakable thing happened to him: someone cut him off and barged in in front of him. (The fellow turned out to be a travelling aluminum cookware salesman.) Leo was so incensed that he couldn't wait to get on the ferry to tell this fellow off. When they got to jawing at each other, Leo found that he "got no respect" at all from this guy, and after they left the ferry he continued to seethe as they drove down the road. Finally Leo overtook the fellow and waved him off the road at a wide spot. He got out, hauled the other fellow out of his car, and got in the last word by cuffing him a couple. Meanwhile, a friend who was a passenger in Leo's car was yelling, "Leo, Leo, don't be crazy!"

The salesman didn't think that this was fair at all so he had Leo charged with assault.

When the case was heard, the magistrate listened to the evidence and, being a North Shore resident himself, decided that since the salesman had, after all, cut in, Leo's actions were not unreasonable. He dismissed the case.

— * * * —

We went back to Nelson in 1972 to a centennial celebration. One of the major attractions of the week was a heavyweight fight between George Chuvalo and a palooka with a long list of spurious-sounding credits. We had ringside seats and had barely settled into them when the bell went to start the fight.

Early in the first round, the other guy hit Chuvalo a hard lick. George was visibly irritated at this so he threw away his script and laid the guy out with two or three thunderous blows.

This episode may have been prophetic because KFP and Nelson were on the receiving end of a knockout punch a few years later, though they hardly qualified as palookas.

— * * * —

We had a master mechanic in the plant who had considerable native talent which he unfortunately mixed with a considerable lack of personal polish. He built some ingenious devices that always worked because he could see things so clearly in his head. His drawings were often only sketched on bits of paper or on the inside of cigarette packages.

The mechanic and I were once asked to go to Boundary Sawmills at Midway to give our opinion on some mechanical problems they were having. My wife accompanied us and we were all invited to spend the weekend at the home of their superintendent, Pete Sherbinin, who, with his gracious wife, made us very welcome.

Gerry and I got up first on Saturday morning and shortly we heard our friend stir. He got up, sat on the edge of the bed, and lit a cigarette. He then proceeded to cough like a man dying. "Ah," he said, "that's a good cigarette." He then came to the breakfast table, shirtless and in his bare feet.

— * * * —

Everyone who has ever operated a mill is visited regularly by vagrant thoughts about piecework. The more you do, the more you make, and all that stuff.

At KFP we made grain doors for the railroads. These were liners that fitted together inside the regular boxcar doors to keep the grain from seeping out. In due course, all this was replaced by modern hopper cars.

These grain doors were all hand-nailed and the nailers were piece-workers and prima donnas. All the components were cut for them. They fetched and carried some of the smaller pieces, but most of what they needed was placed by machine right at their work station. They assembled the unit, clamped it all together in a jig, and then nailed it with thin wire clinch-point nails that went through the wood, struck a steel plate, and curled over so that the unit couldn't come apart. Some of these nails were 3 1/2 inches long and driving them through tough hemlock was a neat trick. They used shingler's hatchets which they weighted up until they weighed about two pounds. They allowed themselves two blows per nail — one to start it and the other to drive it home and clinch the point. An amateur would take six blows and fold up as many as he drove straight.

We could use any species except cedar for these doors, but naturally the nailers preferred anything soft. The softest species we had was white fir, and the nailers could see a forklift load of it coming a mile away. As soon as a load of this material came around the corner, these guys would fight like animals to get the lion's share. They would lift prodigious armfulls of the stock and stagger back to their tables and then run to get some more.

These guys were maddening. They would work like crazy until they had made what they thought was enough money for the day, only $35 or $40 at that time, and then they would want to go downtown for a beer. Because it screwed up the efficiency of the pre-processing, the carloading, the forklift work, and other supporting activities, I had to fight with them to stay for eight hours and make more money.

— * * * —

There were two apartments above the office at KFP. One was kept as a guest apartment, and the other was a social lounge and the home of the "End Of A Lousy Week Club." Late in the afternoon of most Fridays, someone was sure to say, "Do you want to go upstairs and have a couple before you go home?" Ostensibly this was a time for "The Week in Review," but predictably it would run on longer than prudence dictated. Eventually the phone would ring and an irate voice would demand, "Is Ray there?" The answer was always, "No, I don't see him."

Our next words would be, "Ray, you'd better get home.

Maisie's on the warpath." This is not a recommended procedure.

— * * * —

Planermen, by definition, are ornery. At KFP we had a non-practicing alcoholic as a planerman. We were very good friends socially.

This fellow also gave up smoking when he gave up booze, but he did chew. He chewed in the house and spit in the flowerpots, and when we played cards, which was often and usually pinochle, he taped a paper bag to the arm of his chair and used it as a spitoon. You're appalled? Hockey and baseball players are paid a million dollars a year to spit in front of twenty to eighty thousand live witnesses and ten million more on TV. This fellow's wife was a great lady and a ringleader in the movement for mixed company parties.

— * * * —

We had our rowdies and repenters. One of the other planermen at that mill got into a weekend pub scuffle and struck his head on a tile washroom floor. He came in to work on Monday morning but went home at noon because he had a headache he just couldn't shake. His doctor put his finger on the cause: he had a fractured skull.

Another planerman was changing the top head knives in a twelve-knife planer and had all his knives set in place, finger tight. He decided to break off and have a sandwich, but when he returned he completely forgot that he had not tightened down his knives. He started the head, and was chilled by the sound of breaking steel that sounded like breaking bottles. There was steel embedded in the beams above the planer, there were holes in the blow pipes, and the bed plates were all scarred and chewed. He sat and thought about this for a minute before he made his next move, which was to call his boss and confess.

It was while I was at KFP that I got a look at the dark side of union obstinacy. The company was installing a third boiler in its steam plant and it was necessary that it be positioned between the two existing boilers. This meant that the new equipment was located in the centre of the boiler room and this posed a problem for raising the smoke stack that would sit on top.

The supporting structure and the boiler itself were installed by the company's steam plant engineers and plant mechanics, and we thought we had had a stroke of good fortune when we heard that a Vancouver-based crane was in town to erect the steel for a new post office. This machine had a one-hundred-foot-long boom with a jib in addition, and could have handled our job with ease. We approached them and their office O.K.'d the job, but when the crane crew came to take a look, they asked who had done the steel work up to that point. When we told them, they announced that we had been doing union work, and, being good steelworkers, they refused to help us.

We wanted to get on with the job since we were so close to finishing, but thought that if one union outfit would turn us down, another might do the same. We approached the CPR Bridge and Building crew, whose rules were less severe, and persuaded them to come in with their rail-mounted crane. This was practical because the track ran right through the centre of the plant and within thirty feet of the boiler house wall. The catch in this was that, though the B&B crane was enormously powerful and could lift a locomotive, it only had a fifty-five foot boom.

We decided to give it a try anyway, and the dispatcher allocated us a clear-time of two hours between trains to do the job. The hundred-and-ten-foot long stack appeared to go into place without a hitch. The bottom end was positioned on the boiler and all that was needed was to stand it erect, in about the way that native people raise a totem pole. This we could almost do, but not quite. Try as we might, we could not get it upright enough so our small winches could take over and pull it up into its final position. At one point a steel band on the stack that the crane sling was attached to slipped up the stack six feet, causing the top to whip like a fishing pole before it caught itself and settled down. This required that we send up a welder in a bos'n's chair to weld the band in place.

With much straining and the application of body English, and eventually even the desecration of the CPR's magnificent crane by the welding of a piece of key steel to the brake drum to overcome a slipping brake, time ran out and we had to admit defeat.

However, at the last minute, we made a desperate decision. We stood the stack on end on a pad just inside the boiler house wall. The following morning we walked that hundred-and-ten-foot stack sideways a distance of fifty feet, on end. It wobbled somewhat like a broomstick being balanced on a boy's fingertips. By paying out on some winches and taking up on others, we got the stack where we wanted it, raised it ten feet from the boiler room floor, and lowered it neatly onto the smoke box of the boiler.

It couldn't be done, but since we didn't know this, we pulled it off with only minor heart attacks. That was organized labour, but not the steelworkers' kind.

— * * * —

One man's meat....

Clear lumber is the most valuable but the least plentiful grade on Interior logs. Under these circumstances, even one piece would be welcome. Coming from inland species, they aren't even called "clears"; in a continuation of an old pine tradition, they are called "selects."

On the KFP green chain, selects were piled at a place on the chain where, because of their scarcity, the puller had several other odd ball loads to service as well, but even with that he had an easy job.

On one occasion, the loggers ran into a patch of beautiful old growth white pine logs, some of which were four feet in diameter. These all arrived at the mill together in a boom and soon the lumber started to come out of the mill and down the chain. After fifteen minutes or so I looked in at the chain, where I found the clear puller stripped down to his undershirt and dripping with sweat. His explosive comment was, "Nothing but effing selects!"

— * * * —

Interspersed through my stories are good lumbermen, even great lumbermen, who, like some great ball players, had an antic side to their private lives that identified them as clearly as did their professional skills.

Ray Wallace had a "game face" that was his trade mark. He was so organized and well-informed that customers lined up to buy his lumber. They knew that whatever he promised would surely happen. If there was any threat to his reputation from our end, the rest of us felt his wrath and we jumped a little higher and faster to avoid it.

His marriage, on the other hand, was one of those strange cases of opposites attracting, and since they told the stories on themselves, their lives were sort of in the public domain.

Maisie was tall, strident, and excitable; Ray was short and ornery. The result was often a wild adventure, but they always stopped short of anything terminal to their marriage or their lives.

They once invited another couple in for an evening of bridge but, almost predictably, Ray lingered with friends and screwed up Maisie's supper plans. When he called home, he was told that if he wanted anything to eat he should pick up a TV dinner. This he did, but when it was heated and placed before him he looked at it and said, "What's this? You know I don't like turkey." The

cooking half of the team reached across the table, picked up the dinner, and fired it at him. He ducked. It hit the wall and, for the most part, stayed there like a piece of modern art. And it stayed there through the evening of bridge as well, without a word of explanation to the guests.

In another heated moment, one said, "Okay, pack your own clothes; I'm not your mother!" The clothes packer then ventured out on a trip with one pair of socks — royal purple — something that goes with everything.

One sultry summer afternoon Ray took an unannounced cruise on his air mattress, eventually docking

on the far side of the lake, where he found a group of convivial friends and spent the afternoon drinking beer. While he was thus occupied, he didn't think a whole lot about a commitment he had made to take his mother-in-law to the airport in Castlegar. When the familiar phone search was finally successful, he was brought back by power boat to face the music. The music started when the boat was barely half-way across the lake and the words of the song were something about fifty ways to kill an S.O.B.

Ray's den was a dark, windowless, airless basement dungeon where there was a TV set and a couple of pieces of overstuffed furniture with springs poking through the covering here and there. No doubt fixing it up was one of those things he was going to do soon, but there was never any explanation or excuse offered. Many a dapper Yankee yuppie sat down there and painfully endured an assault on his every sensitivity, but as each of them squirmed, I'm sure Ray hid a sly smile.

At one time Ray decided to paint the steps that went down to this room. He painted every other step, intending to go back and paint the others when the first ones were dry. But something came up and they just continued on indefinitely in their checkered state.

As I said, there was no malice intended; it just looked like it. There was the matter of the golf lesson where the classic swing was demonstrated — head down, eye on the ball, left arm straight and a nice easy follow-through that hit the student, Maisie, in the middle of the forehead and dropped her unconscious in her tracks.

— * * * —

When I first went to Nelson, the W.W. Powell Company was still making match blocks. This is a case of a totally unique industry disappearing without a trace. W.W. Powell bought white pine lumber all over the Kootenay area. Their method was to make personal visits to the mills in the area, look at their logs and lumber quality, and give them orders for the grades and sizes they could use. When this lumber was cut, it was piled into air dry piles. A small area on a board was smoothed with a hand axe and the pile was labeled, "W.W. Powell Co." and dated. The mill was then given an advance payment on the lumber in the dry piles.

When the lumber was dry, it came to Powell's Nelson plant where stock suitable for making match blocks was selected. The balance went to regular lumber uses. To qualify as match block material, the wood had to be free of pitch and blue stain; it had to be straight-grained and the knots had to be limited in number and well-spaced.

All widths of this stock were first trimmed in a gang trim to four-foot lengths, and then trimmed again in another machine that cross-cut them into sections about two inches long.

This material went into a room where there was a square circle, so to speak, with four inclining belts, each dropping material onto the next one. Standing around these belts were several women, each with a work-bench on which she had a lead chopping block shaped like an upside-down loaf pan, and a small, very sharp, hand axe.

These women would spend the day reaching stock from the belts and chopping out clear straight grained wood in the biggest pieces they could recover. The good material was thrown down one hole and the knots and unsuitable pieces went down another. The match blocks were then loaded in loose bulk in boxcars and shipped to the Eddy Match Company in Pembroke, Ontario, to be made into wooden matches.

The use of the waste part of this operation is the exclusive knowledge and experience of old-time Nelson residents. It could have been sold as solidified gasoline, but instead it was sold as fuel for heating homes. Most pieces contained a knot or a pitch pocket or a solid pitch streak, and once the stuff took fire in your furnace or heater there was no holding it back. It burned with an intense, uncontrollable flame, and was really a nuisance to use because it burned away so quickly. It was also a pain to put in your basement because there were a zillion pieces in a truckload.

Going back to Fraser Mills for a moment, we occasionally sawed Sitka spruce there. As I recall it, these were always very low grade logs, and one of the products was 4x4 sold to a company who made the plugs that went into the ends of linoleum rolls and other rolled products. Because of the type of wood, there were lots of loose black knots that fell out whenever the wood was handled roughly. These fell out along the length of the green chain.

An elderly clean-up man with a bit of a quirk collected these knots for the same purpose as the reject pine from Powell's. He spent hours painstakingly picking them up and stuffing them into burlap sacks and lugging them home. I guess both these items would work in an airtight heater, but I would be concerned about the heater melting.

W.W. Powell had no burner, and the pile of shavings from their conventional planing operation was a landmark in full view of downtown Nelson. One year we had a particularly cold winter. The lake froze and stayed frozen. The mill stopped sawing, but we tried to continue kiln drying and planing to move out the inventory. The shavings from the planing mill would not keep the boilers going and it looked like we would be completely shut down.

Someone got the idea of trucking shavings from Powell's shaving pile to our plant, so we organized trucks and got going. The front of the pile worked fine but when we got farther back into the pile, the shavings took on a dead black look. When we put them into the boiler fires, they put the fires out. The stuff just sat there and steamed a little. It was one step away from being humus.

Harry Horton was Powell's manager and he played the role of a rube for whatever devilish reason, since he was anything but. I remember a refined and unsuspecting lady telling of sitting next to Harry at a dinner and having him say, when dinner was over, "I hope I didn't splash you too much."

— * * * —

Spokane was the city for Nelsonites, and we made many trips there to shop and to eat great meals at the Ridpath and at the Davenport, with its magnificent wood finished lobby and diningroom.

There was a well-known wholesaler who had the reputation of being a tightwad and a check-dodger. One evening, Ray Wallace and I met this fellow at the Ridpath and went to dinner. He affected the role of the outdoorsman and came to dinner in this elegant room in leather boots and a checkered shirt, into which he tucked his napkin before polishing off a hearty meal. Probably from years of practice, he knew when the waiter was just around the corner with the bill because he excused himself and left the table just long enough to let us take care of it.

The following day we drove to his home, which was farther south, and when we got into the small town where he lived, he asked Ray what he drank. "Scotch," said Ray, without hesitation. "By golly, I think I'm out of Scotch. I guess we'd better stop here at the liquor store and pick up a bottle." With this he looked into his wallet with concern and said, "You know, Ray, I don't think I have enough money with me. Will you take care of it? I'll square up with you when we get home." We did get one drink, but dinner was served almost immediately. It consisted of a casserole of shell pasta with shrimp, something to which Ray took exception and declined to eat.

We left the following morning, but neither the bottle nor its price were mentioned again.

We were once at this fellow's place over a weekend and Gerry was taking in the sun on the patio. A young skunk ran up and made a bed for himself on Gerry's lap.

She yelped, "Say, I hope this thing has been adjusted." "Well, no, he hasn't," she was informed, "I took him to the vet but he wanted fifteen dollars to fix him. I decided not to do it."

Grace was said before meals in this house and it consisted of an informal conversation with the Lord. It started "Well, Lord, we had a pretty good day today...."

One of the enchanted moments of my life happened in Nelson. Since it is such a mountainous place, most yards are on a slope. Ours was divided by a four-foot high stone wall that terraced the back yard, and at the end of the yard another wall separated us from our neighbours.

It was winter and our cocker spaniel was spending his nights in a glassed-in back porch. For a couple of nights we had heard him barking and grumbling, so the next night, when he started again, I got up to see what was going on. In the yard, which was lit by a brilliant clear moon, were two big mule deer that were pawing

through the snow under an apple tree, looking for windfall apples. They were no more than ten feet away from the window. Suddenly they sensed that I was there, and, in an instant, with effortless grace they leaped over the wall. In four or five bounds they cleared the yard and the wall at the back and disappeared.

It was thrilling and it left me with my heart pounding.

— * * * —

At one time in the early 1900's the CPR had track that, in combination with lake barges, took trains all the way up through the Lardeau country to Trout Lake City, which gets pretty close to Revelstoke. Gerry and I played hookey one week and took a holiday right in our own back yard. We drove up through the ghostly remains along the route through Kaslo, Meadow Creek, Lardeau, Howser, Gold Hill, Poplar Creek, Gerrard, Trout Lake City, and Beaton. All of them were gold rush towns that bloomed for ten years or so and then shrank back to hamlets or were abandoned.

A lot of that railroad right-of-way became public road and with the easy railroad grades KFP's logging trucks could haul full loads to the dump at Kaslo, driving at highway speeds.

We felt like intruders when we went in the open door of a deserted building in Poplar Creek and found magazines dated 1934 in perfect condition, except for mouse droppings which brushed off. The last tenant, we found out, had gotten old and his family didn't want him to live there alone, so they moved him to a home. He must have just closed the door and left because there were still a lot of personal things in the house.

The streets of Poplar Creek were long grown over, but here and there in the bush you could see fire hydrants that had been put in when the residents were sure their town was there to stay.

When we were in Nelson, the Miner's Hotel at Gold Hill was still intact. It later collapsed under the weight of heavy snow. This was an ancient building with a squared- off Wild West false front. Only a few vestiges of paint remained, but you could still make out the name on the grey weathered wood.

The widow of the last owner continued to live on, alone, in this little hotel miles from civilization. Half a mile from the hotel, the last section foreman of the defunct railroad also continued to live in a small cabin. Whether they had a grudge of long standing or just hadn't ever had a reason to be friends, each continued to act as though the other was not there; that is, they didn't even speak, let alone visit.

— * * * —

xxv) Kootenay Forest Products, Nelson, B.C., 1957.

xxvi) Carriage at Kootenay Forest Products. Gun shot feed and fast sawyer once sawed 1,000 small short logs in an 8 hour shift making cants for two resaws.

xxvii) Small mill with a big heart. 7′ band head rig, 6′ band resaw, 54″ band linebar resaw. 4-saw solid tooth board edger fed at 400 feet/minute. Production was 1″ boards in spruce, pine, and cedar. 2″ dimension in fir, larch, hemlock, and white fir.

xxviii) One of two Woods motorized 12-knife 6-head planers.

xxix) Vern Vance (L), Ray Wallace (Centre) and the author (R).

xxx) 5′ diameter white pine tree soon to be a log.
In the Lardeau country, Kootenay Lake, 1955.

xxxi) Mary Harvie with a 20 lb. rainbow.

Chapter 7

Up The Fraser River

In 1957, while on vacation at the coast, an innocent conversation with a friend led me to an interview with Stan Laidman at the B.C. Manufacturing plant in New Westminster which was then part of Powell River Company. That old paper company made a brief foray into lumber production and sales between 1950 and 1959, before the Powell River Company was merged with MacMillan Bloedel.

Stan was in the process of doubling the production of that big old mill, and he had so many details in the air at one time that it was decreed that he should get an assistant to look after production while he was tied up with other matters.

Stan was a man so obsessed with detail that he could have respectably died from the strain when he was forty-five. The saying was that he knew how many threads there were on every bolt in the plant. He was still alive and alert at the age of eighty-five when this was written.

As much as I loved KFP and Nelson, Stan still persuaded me to make the move, and I started another round of new experiences with new people.

The Powell River Company was bigger than any company I had worked for up to this point, but it retained a small company relationship with its employees. Harold and Joe Foley were men of exemplary character and were a credit to the industry.

Both Harold and Joe knew most employees on a first name basis and kept in touch with their general welfare. At the time I joined the company, Harold was the chairman of the B.C. Cancer Society. If someone in the company did develop the disease, Harold personally saw to it that the latest medical knowledge, equipment and procedures were made available without any financial concerns for the employee's family.

John Leirsch was the chief executive officer and he left a gem of wisdom for posterity. To the suggestion that we should build some inventory to catch a strengthening market in the Spring, his response was, "Look, you sell today's production today and tomorrow's production tomorrow. If you want to gamble, put your money in the stock market."

— * * * —

BCM and the Westminster Shook Mill were the major Powell River Lumber Co. mills. At the time of the purchase by Powell River Company, they were both owned by the McDonald family and, having grown up in Queensborough, I was quite familiar with them. I was also familiar with the McDonalds, though mostly from the prejudiced reports of the local citizen/employees who thought they were hard-boiled operators.

Bob McDonald ran the Shook Mill, and he was a familiar sight driving one of those huge 1942 Chrysler coupes down Ewen Avenue. He drove this car to work, and since he chewed constantly, he spent a lot of time spitting out the window. Time had built up a thick crust of tobacco on the door and he seemed unconcerned about removing it. Errol Wintemute later bought the car as a museum piece, but he didn't preserve the tobacco.

The old BCM office was built beside the original Queensborough Bridge and entry was off the sidewalk of the bridge itself. When I started wholesaling, I went in to see Ian McDonald, and in an offhand manner he advised me that they weren't taking on any new customers right then. Now that's what I would call comfortable.

Before it became a cedar mill, BCM cut hemlock for the export and domestic wood box business and when I arrived at the plant, much of the old box factory was still intact. The stitchers and nailers and printers were gone, but the Lindermann jointer and cutoffs and cleat saws and rip saws and the twin band resaw were still there. With this apparatus, they cut lumber up into millions of short pieces of exact sizes for assembly into boxes — otherwise known as box shook.

BCM also peeled green fir veneer to make berry boxes. When I was a kid, I would go there to buy strawberry boxes for our little farm. There was a tricky box which was collapsible. It was stitched with wire and scored in the right places so that with a squeeze and your mouth held right, the thing would snap open into a perfect little box.

Sometimes they would be out of these collapsed boxes and I would have to take made-up boxes that were intended to be shipped in crates. I would ride my bike the mile or so back to our place with a couple of hundred boxes strung on wire. I was so completely enveloped that I went down the road looking like a big bunch of boxes on wheels.

— * * * —

When you arrive at an old established mill like BCM, you feel inadequate because so much has gone on before and everybody knows all about it but you. Greenhorns were not a problem; the largest percentage of the employees were old-timers and their ways were set. At a later time, when the plant had become part of MB, Jack Forrest blurted out in exasperation, "Jeeze, you guys never even joined Powell River, let alone MB."

— * * * —

For all Stan Laidman's reported toughness, he was really a very nice guy who had a soft spot in his heart for kids, dogs and cooperative employees. He was quite wrung out one time when he found that one of the boilerhouse cats, being a little too inquisitive, had been caught in one of the dry kilns when the doors were closed and the charge started. A normal schedule could be five days with live steam controlling the humidity and temperatures running up to 190 degrees at the end of the schedule. This poor little cat survived this ordeal, presumably by staying at the floor level and near the doors where the temperature would be the lowest. She was a bedraggled looking bit of dehydrated cat when she was rescued.

Stan personally supplied the catfood for this lot and fed them every day and on weekends on his walks around the plant. On the other hand, he could see the workers in the planermill by peeking through the lumber piles in the dry shed, and if he thought someone was slacking, he was not above collecting his own evidence.

— * * * —

It was at BCM that we took a lesson in the explosiveness of things that aren't gunpowder. Since the plant processed a large amount of dry lumber, the sawdust and shavings that went into the fuel bin at the boiler house were like tinder. One day, we got just the right combination of air-suspended dust, oxygen, and a spark from an unknown source, and the place went off like a bomb.

We were sitting in a meeting in a nearby yard office and went through that eerie feeling of vacuum, then concussion that is part of this kind of explosion. As the bin went bang, we fell over ourselves getting outside to see what was going on. When we got out, there were still sheets of corrugated iron in the air. Miraculously, there were no fatalities. The worst injury was to a man who was hit by a sheet of flame where he was sitting shirtless on a railing outside the building on his break.

As soon as everyone realized what had happened, Harold Renwick got the supervisory staff together. He cautioned us that the place would soon be swarming with reporters and that until we knew the seriousness of the injuries and the cause of the accident, we should just refrain from saying anything.

This wouldn't do for the reporters who needed a story. They found their informant in a guy who had been working inside the sawmill building some distance from the boiler house, quite removed from the scene. He was only too happy to give them his theories on the cause and the effect of the explosion, and this was duly reported in the press.

— * * * —

BCM was another plant where there were two lovely

old horizontal steam engines left in perfect condition from earlier days. I saw them run when I was taken through the plant in 1941 by my erstwhile grading mentor, Mac Ewart, who had gone there as a foreman.

— * * * —

Occasionally, the production manager gets thrown a sticky problem. There was a sprawling wood frame plant across the street from BCM. It was originally Heaps Engineering, a company with a colourful history that went back to World War I and before, but it had become AMFAB, a manufacturer of glulam beams. These buildings were not sprinklered and we lived in constant anxiety because of the sparks that blew over their way out of the top of our refuse burner.

One night we had a real gale blowing in the direction of AMFAB, and I had choked the burner drafts down until the fire was just banked. As we continued to produce, the burner gradually filled to the top with slabs and waste.

Eventually the mill lost some production but the burner would not be denied.

We had a fire that was so hot that the entire burner was red and shuddering, and Blake Morton and Colin Hart from AMFAB kept phoning and saying, "If you burn our place down, we'll sue you." Ah, yes, I remember it well.

— * * * —

There were some blissful moments. Once every month or so I would go out with Gary Galbraithe or Jimmy Chapman from the log department to look at logs that were being offered by brokers or as trades with other lumber companies. We would either drive to Horseshoe Bay and rent a water taxi for the day or take a float plane from the Fraser River, in either case armed with a box lunch and a case of beer. We would then cruise or taxi the plane up and down Howe Sound, walking over and approving various booms. Later on, if one of them turned out to be a dud, someone was sure to say, "Who bought this crap?"

There were certain things that were dependable clues to log quality, but I will pass along the following story without explanation. Those who appreciate it can have a smile with me:

At one time, hockey legend Babe Pratt was employed by McDonald Cedar, and they trained him to be a log buyer. When he came back from an early outing, they asked him, "...and what about the knots?" "I didn't see any knots," Babe said. "If there were any, they must have all been on the inside."

— * * * —

BCM had a reciprocating splitter saw that they used to halve and quarter oversize logs so they could get them into the mill. This big saw stroked with the motion of a carpenter's hand saw but the blade measured twenty inches from the back to the teeth and was twenty feet long. We also bucked off flared butts with a drag saw. These were bought by a couple of guys who made shakes out of them. I wonder how many times they tempted the boom men to trim those butts just one shake block longer.

— * * * —

When I ran the BCM plant, we were in one of those situations where cost and profit were centred wherever the company decreed that it should be. We paid the logging division a theoretical market price for our cedar logs and this and other circumstances resulted in us not only producing red cedar but also red ink.

The division never did make money, but curiously, at the bottom of each operating statement there was a line that said, "Contribution to overall company profit and loss..." and it was always a black figure. But the red ink coloured your thinking, and it was a bit depressing.

One day Harold Renwick, that super gentleman, put things in perspective. He said, "Steve, don't run a haywire operation. If something needs to be fixed, fix it right. We're not that short of money."

— * * * —

I found that common sense was still working when engineer Chuck Herstrom debunked my contention that the way to make an ultimate choice was by comparing pairs (of alternatives) and choosing the better one until there was only one left. A second savant agreed that looking back would only be making choices between second choices. The key is to not look back.

— * * * —

One of the most interesting character studies for me is the guy who plays the part of the know-it-all blowhard who gets under everyone's skin and then, by golly, comes up with a performance that proves he is as good as he says he is. If he's got the real estate deal of the year, you can't help hoping that his basement will develop an artesian well.

We had one of these know-alls at BCM who went the wrong route to ever make it with Laidman. He got to be a shop steward and worked his butt off trying to dig up circumstances that might be remotely interpreted as transgressions against the contract. This, along with his infernal boasting, didn't exactly set him up as the man of the year around the office.

One of many things he claimed to be was a sawyer, and when the mill revision was done, he was given a chance to show his stuff on a new double-cut pony rig, cutting small merch (tight knotted) logs.

Trapped in a sawyer's box where he couldn't talk to anyone, he proceeded to saw rings around the other sawyer. Not only that, he had the skill and tenacity to do it consistently for eight hour shifts under production pressure.

When BCM was closed, the crew was dispersed, with many of the men going to the Canadian White Pine Division of MB. Sid Henry, a lifelong family friend, was a mill foreman at "the Pine" and he fell heir to my bumptious friend.

I met Sid on the street and he said "Jesus, do all of these guys talk and none of them listen?" I knew exactly who he was talking about.

— * * * —

When Stan Laidman hired me, he revealed his entire work ethic in two sentences. "Steve, remember that this is a two-shift operation. I don't expect you to come back here every night, but most nights you should look in." Most nights I did, but the night I'd miss was always the night there would be an incident. The following morning, Stan would say, "Well, what did you think of that last night?", to which I would have to reply, "Think of what, Stan."

— * * * —

One day we were all caught up in a well-publicized move. We were merged with MB, and some of the fun went out of our little Powell River family. Ken Bradley, Mike Lepage, Erroll Wintemute and Bob Thom all chose to leave quickly. A lot of others didn't get a choice. A major takeover is a scary thing.

The first Christmas after the merger, a group of Powell River Lumber people got together for lunch and started a tradition that has lasted for thirty years. Some members of the group continued to work for MB until they retired (one still works for the company), but all of them retained an outlook on life that was a little different because they had seen life outside of MB. Most of them were looking back on only one other employer, or two at the most, so their outlook was quite different from mine since I had worked at such a variety of places.

When the rumour went through MB that there might be some merging of Powell River executives into the new organization, J.V. Clyne assured a meeting of MB people that "It was never our intention that the tail should wag the dog," and subsequent events showed that he wasn't kidding. Nevertheless, I stayed on for another three years as an MB man, eventually working in sales.

No one as big and successful as MB could be wrong about too many things, but it was a little galling to be told that you should forget anything you had done before because your only concern now should be to learn to do it their way. Of course they were right, and if there was any problem between MB and me, it was my fault, not theirs.

Their rigid disciplines had a noble source. Jay Potter told me a story that only an MB insider could appreciate. He was accompanying a tour through the Canadian White Pine Mill and was asked a question by H.R. MacMillan himself. "Gee, I don't know the answer, Mr. MacMillan," he replied, "but I'll find out and let you know."

Someone who overheard this said, "Gee, Jay, you made yourself look kind of dumb. Shouldn't you have told him something?" Jay said, "If I had, you can be sure that sometime, even if it was six months from now, H.R. would call me up and say, "Jay, last year you told me something about this and, you know, you were wrong." Apparently H.R. seldom forgot anything and the whole company seemed to sense his constant presence.

Jim Lawson took me aside one day and said, "Steve, you're going to have to learn to think in numbers. If you can't do it, you probably won't make it."

— * * * —

Len Kingman came on the scene as the new manager's "assistant to" which, in my apprehensive state, I interpreted as "spy" and "finger pointer," but eventually we worked together in sales and became good friends.

Once a month Len and I got a forecast of what our log diet would be for the following month, and from this we prepared a sales forecast based on historical recoveries from similar logs. These assumptions were broken down into minutely detailed estimates of size, grade, and length recoveries — percentages of percentages of percentages. Our best guess was then shaken and stirred with the best guesses of all the other divisions, and the result was a forecast the likes of which was seldom seen. Ken McIntyre received this information, but I'm not sure what he did with it.

It was one of these meetings that gave birth to a story that is part of MB sales legend.

Gene Squarebriggs, known to one and all as "Square," was the sales manager at Port Alberni. He was also an eccentric extrovert. He was so irreverent, so often, that people just let him go, and tried to follow the thread of facts that he wove in with the B.S.

At one of these sales meetings where there was some senior brass in attendance, Square got ready to give his report. He reached into his briefcase and took out a comic

book from which he proceeded to read out his heroic volumes of facts and figures.

— * * * —

Ernie Shorter was a practical sawmill man who rose to be president of MB. He was of the old school and deplored the arrival of calculators (they make your mind lazy), let alone computers. One shouldn't laugh because people like Ernie, Len Stone at Fraser Mills, Carl and Ernie Lillos, Manny Bouthot, and countless others were quick and accurate, without using calculators, to a degree that will never be seen again. Most of them could run up multiple columns of figures combining groups as they went, adding two and three columns at a time.

Ernie left a cliché for posterity. He made regular tours of all the mills and never failed to find a piece of lumber lying on the roadway where it had fallen off a load. He would say to some groaning foreman, "If that was a $10 bill, would you leave it there or would you pick it up?" I don't recall more than one answer to that question, and it was always sheepish.

Jack Forrest was a member of a prestigious committee that was chaired by the president and contained some of the bright lights of the company. One day Jack had another important committment that conflicted with a meeting of this committee and so he delegated me to attend in his place. As bad luck would have it, I got caught in a traffic jam behind an accident and arrived at the meeting quite late. I paced up and down the hall for a minute wondering if I should die, quit, run away, or open the door and walk in. I chose the last one, and the president said, "Well, I guess we won't start again for you, Steve."

— * * * —

In the end, they demolished Stan's newly rebuilt mill and took the major components to Vancouver Island. Meanwhile I moved into sales and got a little better feel of things working with Squarebriggs, Lawson, Harris, Montgomery, Dalton, Prior, and other free spirits who knew how the system worked.

There is no doubt that there was a bond between the MB veterans, but there was also a mystique that shut you out until you could prove you belonged — and could remember never to say, "When I was at...."

— * * * —

I now realized that I would never be a big company man. Trying to live my life as though I was filling out a form made me uncomfortable. I found I was more concerned with how I was doing things than I was with what I was doing. And nothing could replace the satisfaction of working for a small company, where major decisions were made right on the premises.

The only way my system could work was when I called the shots and lived with my mistakes. As soon as I started worrying about making mistakes or making less than perfect decisions and having to explain myself, I was finished. They called it "being defensive." I found it intimidating, and it destroyed any usefulness I might have had for the company.

The day I left MB, I had been called downtown to the head office personnel department to explain why I was leaving. On the way out of the building, the elevator stopped and J.V. Clyne stepped in. There were just the two of us, and J.V. asked, "Do you work for us?" "I did," I replied, "but I am leaving the Company this morning." Needless to say, we completed the trip in silence.

— * * * —

If I sound a little jaded about MB, I should try to correct that impression here. We were mismatched and there was nothing more to it than that. I never met an MB production man who didn't know what he was doing. Of course they had the best of logs and good equipment, but they certainly didn't leave anything on the table through inattention. This didn't always gladden the heart of a remanufacturer who was trying to eke out a living by gleaning.

People make life pleasant wherever you work and MB was no exception. In spite of what you might think of the system, there was always a mitigating surface condition, fostered by kindred spirits, that kept you up and running, and I went on later to become one of MB's bigger customers, dealing daily with people I had worked with when I was with the company, so there were no lingering feelings either way.

— * * * —

xxxii) The Author and wife, Gerry, 1959.

xxxiii) Powell River Lumber Company stalwarts. Twenty-eight years of nostalgic Christmas lunches. Top row, from L to R: Bob Thom, Tom Hanna, Ed Dunkinson, the author, Mike Lepage, Tony Pare. Bottom row, L to R: Ken Bradley, Fred Ridley, Errol Wintemute.

Chapter 8

Two Lost Weekends

When I left MB in 1963, I took a one-shot assignment with DeCew Lumber, an old established Vancouver wholesaler. My objective was to assess the changes in sales philosophy and practices at the mills, firstly in the light of the rapidly expanding communications facilities and, secondly, with the reduced risks of selling to the large stocking distribution yards in the consuming areas which were well financed, but were neither wholesalers nor retailers.

This distribution revolution was well underway, and the small local wholesaler was left to sell to small, high-credit-risk retailers in far away places, unless there was something unique about his mill connections.

This tour took me a month, and I retraced my steps of several years earlier, visiting practically every mill in the B.C. Interior. I wrote what I thought was a good report and decided to stay on with them, at least until something more substantial came along. There was no commitment.

— * * * —

This job involved a move to Vernon, where we rented

a house on the shores of beautiful Kalamalka Lake. Not bad. We lived in Vernon for a year and had good times.

The kids had a memorable tree house and the dog had Kokanee salmon that spawned and died in a creek that went through our yard. These he rolled in, and then came to the door to be let in to be family. We would bath him and keep him in the house in the hope that he would forget about the fish. The minute our eye was off him, he would go right back and do it again.

This same dog, a corgi, had a serial dog fight with a short-haired, tight skinned, muscular giant of a German pointer. These fights were short, invariably ending with the corgi on his back, showing the whites of his eyes. I was so mean to that pointer and he was so undeserving of it. He never started the fights, but it was hard to love the pointer and still keep your own dog alive, so I would bash him with anything I could reach to drive him off.

— * * * —

The pointer's owner arrived at our door the day we moved in and announced, "Hi, I'm Ed, the neighbour." Ed was a character right out of a Bob Newhart sketch. Ed built a boat in his back yard, a good sized runabout with an inboard engine. As the day for launching came closer, plans for a major event were honed and polished.

One summer day, we took Ed's boat and a picnic lunch and a bottle of champagne down to the lake for the big event. The boat went into the water without a hitch and Ed went with it to start the motor and show off his masterpiece. Ed, unfortunately, had overlooked one of the major parts of the boat, the drain plug. With Ed frantically paddling with his hands, the boat slowly sank to its gunwales. In the meantime, we were caught in one of those "didn't know whether to laugh or cry" situations so we laughed. Then we drank the champagne anyway.

— * * * —

Tosh Yakura ran a planing mill at Coldstream. It was he who had told us that his father had been a Canadian soldier in the First World War. One day he referred to the expulsion, and I said, "Do you mean that they moved you, too?" "Yeah," he said, "I guess it was all the same yellow to them."

Ted Parker was a quick-witted, wise-cracking character who had made a specialty of buying stock for the manufacture of glulam beams. He did in fact come close to having a corner on the supply of this material since he had convinced the beam makers that he was the only one who could get them their material on grade and on time.

Ted drank too much and his verbosity could get a little tiring, but he had a good ear for the language and a wicked sense of word play.

There was a nice but unlettered fellow in the DeCew company who Ted treated with cruel condescension. He told me with glee of how, when the Commonwealth Games were in Vancouver, this fellow had asked him to get some tickets to the bicycle races at the China Creek Velodrome. Ted checked and told the fellow that the only seats left were on the periphery. He then heard his friend reporting to his wife that the only seats Ted could get were on the periphery and "you wouldn't want to sit way down there." Something like that tickled Ted.

— * * * —

It takes a little of everything to round out your life. I tried working with a practising alcoholic. Those who haven't been close to it don't know what the far side of drinking looks like. When this fellow was on the stuff he left the office a dozen times a day with a different excuse each time.

Howard DeCew was a fixture on the local wholesaling scene. He was the second generation in a family business and he was in his late sixties, so it went back a long way. His specialty was white pine. His contacts were so well established that even the major companies sold through him. Howard was the Silver Fox. He smoked big cigars, drank a little, and put on his best smile for the ladies.

Once when we were on a trip to the prairies, Howard took a notion in Winnipeg to get a new hat — a particular brand of hat that he would buy at Hudson's Bay Co. We tried to walk there on a windy, 10 below day. It was practically impossible, but we accomplished it over a period of half an hour by striding a block and then ducking into a store. If it was a big store we went in the first door we came to, then walked all the way through the store and out the far door.

— * * * —

I learned a little more in Saskatchewan from a crusty old gentleman named T.M. Ball. He didn't like some lumber we had shipped him. When I drew on my vast knowledge to explain to him that what he was objecting to was on grade and quite within the rules, he said, "I don't care if it's on grade or not. I don't like it and I want you to take it out of my yard."

I probably hadn't helped matters any when I wrote him a letter in which I meant to suggest that he was a shrewd buyer, but had referred to him as a cheap buyer. He was waiting for me on that one. Howard, the diplomat, assured Mr. Ball that I meant that he "bought close" (Howard's term). I thought he was cheap and tough, but I could have been wrong.

— * * * —

DeCew was a short bus trip. It might have gone somewhere interesting, but it didn't seem too likely that it would, as Howard was close to seventy, close with his money, and pulling in rather than expanding. I got off the DeCew bus and onto one marked Kicking Horse, going to Golden.

It has taken me a long time to put a humorous face on my Golden experiences. It is funny in a perverse way, now that I am standing back far enough to not see the rough edges.

The plant had been a chronic loser for reasons that seemed obvious enough to me at the time. One night the salaried staff all went to dinner at Lake Louise with the officers of the American company that owned us. The dinner was to be a "without prejudice" free-for-all discussion in which everyone would get a chance to air his views. Quite accidentally, I drew a seat beside the president. After we had heard several people assert that five million dollars here, and seven million dollars there, and another million somewhere else, would cure our problems, I said quietly, "Mr. Mudd, as far as I can see, we have an average sawmill, average logs, and we make good lumber. I think that what we should do is quit fiddling with the plant. Leave it alone, let the key people get to know each other a little better, and it will probably turn itself around."

He said, "You're probably right." But I didn't last to make my point, and the confusion continued until the company was sold.

If the plant was floundering, it wasn't the fault of the crew. One night after I had finished my supper, I went back to the mill for my usual look in on the night shift. The temperature was thirty-four below zero and a bitter wind was howling. I went to the men who were working outside on the green chain and asked if they wanted to pack it in and go home. "Oh," they said, "we're here now; we might as well stay."

On first glance, everyone at Kicking Horse, from the janitor up, seemed to be an assassin looking for someone to hit. I made it past my second year (better than average for Kicking Horse) before someone hit me. The Shadow was not the only one who knew what mischief lurked in the hearts of men (and women), and so the less said the better.

The characters on the fringe of this drama, however, were fourteen carat eccentrics and we had our share of laughs, even there. For one thing it was Peyton Place West. There were as many loose boards in fences as there were loose ladies, and enough daring Lotharios to make it all work.

Looking back, it was a modern version of one of those eighteenth century plays where some people keep going out through doors just as other people come in through other doors. The timing is split second, but in the cutaway set, the audience can see that everyone dallies but that no one quite gets caught.

Trust Golden to get it wrong. One night our next-door neighbours seemed to have two men and one lady in the same bedroom. One of the men was the lady's husband, and he was not amused. The other left in a hurry with his pants in his hand.

I should have known that something wasn't quite right in the first week I was there. I got involved in what Jimmy Durante called an "error of mercy".

I stopped on the Rogers Pass Highway late one night to help a very elderly motorist who was stranded and distressed and I decided that the best thing I could do was to get into Golden and try to find a mechanic who would go out to help him.

I guess I exceeded the speed limit because I soon found that a Mountie had followed me into an all-night garage. He couldn't tell a good Samaritan from a mad bomber so I was hauled into court where I was given a condescending lecture and a $75 fine.

— * * * —

The ones in our family who had the most fun in Golden were our two girls, who were now teenagers. Summer in the Rogers Pass is a busy time and since the highway had just opened the year we arrived, every restaurant in town was crowded. Our girls were pressed into service before they had unpacked their bags.

I'm sure every family has giggle sessions, but when Sue and Mavis get going on their Golden stories, they break each other up. Most of the stories are, of course, "you had to be there" stories, but Mavis tells one that has a twist.

The head waitress was an organized lady who insisted that everyone follow her efficient routine. One day she caught Mavis leaving the kitchen empty-handed when she should have brought back a tray of clean dishes. For this Mavis got a blast. Feeling a little miffed, she went back and got some dishes. When she put them under the counter, she may have done it with a little extra emphasis because a cup went flying and broke on the floor. A shard nicked the shin of a lady customer who was sitting on the other side of the counter, drawing blood. Mavis was mortified, and rushed around to the woman to apologize. The lady declined help, took her crutches and hopped off. She only had one leg and Mavis had wounded it.

One of Sue's rough and rowdy logger customers was in court, charged with shooting his neighbour's dog. In the true quirky fashion of Golden justice, the magistrate remanded the case and ordered the defendant to reappear, shaved and properly dressed.

One night a bunch of kids, boys and girls, managed to get an old-fashioned hand-pumped speeder onto the train tracks and made the 10-mile round trip from Golden to Parson and back again in the dark without incident.

There was a delayed news report from my kids on that one — about a five-year delay.

Their cousin visited them from the teeming Bay area of California and chirped, "Gee, what a neat place to live; you can get to everything by just walking." That's one way of describing a small town.

— * * * —

To the pronouncement that nice guys finish last, I say, "Go soak your cliché."

Golden is the site for a testimonial about one of the great gentlemen in the industry, Ward Allen.

At the Golden mill, hemlock was a minor species.

Once a year we sawed about half a million feet of it and put it on strips for several months to air-dry before it was processed. When the time came to plane and ship it, I made my best estimate of the quantities, sizes, lengths and grades that would develop, and put out a list to my most likely customers.

Ward Allen was pretty much a back-to-back wholesaler, which means he didn't speculate. When he had an order in his hand, he bought the material to fill it, and, conversely, when he was offered material by a mill, he waited until he had it sold before he gave the mill an order. But in one instance, while Ward was absent from his office, one of his men took an unauthorized and uncharacteristic flyer and bought my entire offer of hemlock on spec. This took me off the market and prevented me from selling it in bits and pieces to people who knew the stock and had had it before.

The guy's timing was terrible. The market was in one of its periodic swoons and he couldn't get anyone to even make him an offer. As hard as I tried to pry some rail destinations out of him, time dragged on with no action.

One evening I was in Chicago and a phone call came to my hotel room. It was Ward. He said he found himself in a ridiculous situation but he wanted to do the right thing. "Before I put this millstone around my neck," he asked, is there a chance your regular customers would still take their portions, even if they offered prices lower than I'm booked at? If they will," he said, "I will gladly send you a cheque for the difference."

It all worked out without a ruffled feather because Ward is that kind of a fellow.

Later, Ward and I were friendly competitors in the remanufacturing business and we spent many hours commiserating over what the villainous "they" of the moment were doing to us.

— * * * —

One of the pleasant things about Golden was that on nice weekend afternoons, you could get out of town. It was a short drive over into the National Parks, Lake Louise and Banff. One day I got an unsolicited business lesson which I will pass along to you.

We were in a restaurant talking to the proprietor. When he heard that we were from Vancouver, he started to talk about this restaurant and that restaurant, and then said, "A funny thing happened in that one. I went in to see the owner, who is an old friend, and asked him, "How's business?" "Not so hot," he replied, "I'm not making any money." The storyteller said, "I looked at the premises, looked at his bill of fare, which was attractive and varied and well-priced. I looked around and saw a lot of people eating, so I said, "Baloney. You are making money." "Well, I'm not," was the reply. "Well," I said, "you had better have a serious look at what is going on. Who is your cashier?" "Oh, I don't worry about her," he said, "she is my oldest employee." You know the rest. She was working a clever, one-for-you-and-one-for-me split on the cash.

Everybody can use an audit and a knowledgeable friend to take a detached look, just in case something obvious is being overlooked. (We know of an outfit that is fifty thousand times bigger than this, and their outside observer is called the Auditor General, but nobody pays any attention to him.)

— * * * —

xxxiv) Author, right, and Alan Morrow, Kicking Horse Forest Products, Golden, 1963.

xxxv) Log inventory, Kicking Horse Forest Products, 1963.

Chapter 9

Truth and Consequences

These cautionary tales make a strong case for the middle-of-the-road. Somewhere between the stuffed shirts and the profligates lie the people that the world accepts as interestingly flawed. Without them, we wouldn't be able to stay awake.

— * * * —

There is an apocryphal story about the mill that has a customer who is a chronic complainer, continually reporting off-grade material and shortages from the invoiced footage.

The mill becomes suspicious, and, to test the customer's honesty, ships him a carload and mails him the invoice and the bills of lading. However, they also instruct the railroad to delay the car en route, then they wait.

Predictably, the customer calls and complains on the quality of material he couldn't possibly have received, and the cat is out of the bag.

This may have happened once in history, but the story is fondly repeated as the real-life experience of half the old-time sales managers in the industry.

— * * * —

A great story that is probably true is credited to Jack McMillan, who was an early MacMillan sales manager at Port Alberni, though not related to H.R. He once sold two carloads of flooring to a new Winnipeg customer about whom he knew very little. He shipped the first car but decided to wait a while before shipping the second.

Shortly, the customer phoned to ask when he might expect the second car. Jack's reply was, "Well, I thought I would wait until you paid for the first one before I shipped the second."

The customer's response to that was, "Well, you'd better just cancel it then; I can't wait that long."

— * * * —

The world should be a better place when you consider the bad examples we have set before it for its guidance. A wholesaler whom I knew well indulged in a feat of subtle larceny for a while and I was unwittingly involved both in making his scam possible and in his eventual exposure.

Lumber is classified by distinct grades. These grades permit or limit the characteristics which are common to all lumber, regardless of the mill that produces it, and this leads to uniform quality across the industry.

A customer, however, will sometimes ask a mill to supply a high grade with a percentage of a lower grade mixed through the shipment in order to lower the average cost of his purchase. As long as each piece is stamped with an identifying grade mark, there is nothing wrong with this practice.

When attractive appearance is the basis of a grade, grade stamps on the faces are not desirable, but not everyone is equipped to grade mark the ends of the pieces. This is where my scoundrel ex-friend and customer saw his chance to get rich. He bought carload after carload of unmarked mixed grade spruce boards from me until one day I got a call from Dan Broderick in Pittsburgh, asking me, "Steve, just what are you selling this guy?" I told him and he thanked me for the information.

My customer was selling the carloads as straight high grade, and whenever someone questioned the quality he would say, "Gee, the mill must have slipped up," and without argument would immediately give his customer a rebate on his invoice.

For every customer that squawked, several others would either not notice or would just grumble that it wasn't very good stock.

Apart from his being crooked, it wasn't doing our reputation any good, so I quickly cut him off.

Fortunately, these shady characters are outnumbered in a fraternity that prides itself on its ability to complete major transactions by telephone in an atmosphere of mutual trust.

— * * * —

One of the species we sawed at Kootenay Forest Products was hemlock. These trees seemed to struggle to live on the rocky side hills, and much of the lumber they produced had grown-in stresses which sometimes caused it to warp and twist when it was kiln dried.

Sometime in 1955 we got a large order for 2x6 to go to Marlette, Michigan, which is near Saginaw. We processed the first car in the usual way and shipped it, but it wasn't long before the customer was on the phone in a dither, saying that he couldn't use it. I went east to see what it was all about, and what I saw would have been funny if it hadn't been serious.

What we didn't know when we sold it was that the customer was gang ripping the stock to 2x2's for wall framing in the construction of mobile homes. We had made no effort in our drying schedules to relieve stress, nor had we graded out things like compression wood and boxed hearts, two conditions that are notorious problems in dry remanufacture stock.

We went out to their plant and a foreman started up a multiple rip saw. He put a few pieces of our stock through, and what came out was comical. Each piece of 2x2 headed in a different direction, some curving into semi-circles, some turning up and others turning down, but hardly a piece remained straight.

Just to rub it in, they put through a few pieces of soft, straight-grained old-growth coast hemlock from St. Regis in Tacoma. The 2x2's from this stock looked completely relaxed and just laid there, flat and straight. It was embarrassing.

Turning my optimism up to high, I told the fellow I knew how we would have to dry and what types of pieces we would have to exclude to make him a suitable product. Miraculously, the purchasing agent called his superintendent and asked him if he could find a use for our stock, just the way it was, as 2x6, and the answer was "yes."

It was a fairy tale settlement. His explanation was that since we took the trouble to come to see him, he would work with us, and the subject of money never came up.

— * * * —

As noble a fraternity as we are, we are not without our one-of-a-kind characters. One of these, a fellow named Ed, who I ran into again and again through the years, must have thought he was looking at the world through rose-coloured broken glass.

He had been married five times at my last count, and included in a string of personal tragedies were the loss of a child who drowned in a swimming pool; the loss of a construction crew, who were like family to him, in the crash of his company plane; and the loss of two friends who, in Ed's absence, were staying the weekend at his home and were killed when the house burned to the ground.

This has not been the mood of my stories to this point and this was not the usual mood of my friend, who was probably the most resilient man I ever met.

Just after the plane crash, he was in Vancouver for a period of time and told me that he was having some mail directed to our address. When it arrived, it was his bank statement along with his cancelled cheques and the bank's assessment of his financial affairs.

He spread these out on a table and looked at them attentively for a long time. Then he announced that he was going home to Florida. He also advised me that he was momentarily short of cash and would like me to introduce him to my banker so that he might cash a cheque for a few hundred dollars.

Shortly after that, I got a call from my banker saying, "Hey, what kind of friends do you have? This guy's cheque bounced."

I paid off the bank and waited to hear from Ed. Two years later, when I had written him off and put him out of my mind, I got a call from him and, without any chit chat, he said, "How much do I owe you, Steve?" A good cheque arrived in the mail shortly after.

Trouble seemed to follow Ed around. We once shipped him a carload of clear-faced 4x6 decking from BCM. This was for a church roof and it was expensive material. The stock was carefully hand-loaded and shipped in a regular box car. We had no hint that the car might have been previously used to load grain, but when the train got underway, kernels of wheat somehow sifted out of the car lining and onto the lumber. With the vibration of the train's travel, the grain then worked its way down through the car and in between the pieces of lumber. The result was that the faces of many of the pieces were marred with what looked like insect borings. There was so much money involved that Ed stopped unloading and called for someone to fly down to have a look. It was an open and shut case for Ed. I believe we got the railroad to take part of the pill for supplying a dirty car, but the rest was ours to swallow.

There was a side-light to this incident. As soon as I got there, Ed resumed unloading the car. He went down into the little town of Mount Dora and hired three black men off the street. It soon became apparent that he only intended to keep two. After they had worked for an hour, one of the men ate a sandwich that he had brought with him, but he certainly didn't let this interrupt the pace of his work. "Are you hungry?" Ed asked. "A little," the fellow replied good naturedly. "Well, if you can't wait until lunchtime to eat, I guess I don't need you around here," Ed said, and with that he fired him.

The fellow looked so crestfallen that I felt like taking up his cause, but Ed was looking for an excuse and would have told me to mind my business. Still, the way he did it seemed so casually insensitive that I was sure he had done it before, and it made me cringe.

Once, when I was walking along the outside of the Powell River Lumber sales office, Mike Lepage opened the window and reached a phone out to me. "Here," he said, "someone wants to talk to you." I took it and Fred Ridley's distant voice said, "Steve, I looked death right in the eye last night. We turned end-for-end on a dry highway at 90 miles an hour." I said, "You were driving with Ed." "That's right," said Fred.

One night we were at Vero Beach, Florida, and Ed announced that our evening was planned. He had a friend who managed a motel there, a fellow who had retired from a superintendent's job at an automobile plant in Detroit. The plan was that three good ol' boys would get their feet up, have a drink and tell a few stories.

First we had to get some talking lubrication which, in this case, was Cutty Sark. Ed and I also crashed and stumbled through the dark to get to where a pier left the shore and ran out to a small square wharf well out into the ocean.

Even though it was evening and dark, there were people out there, working at cooking, cleaning and cracking snow crab which they were packing in tins — back meat and leg meat separately — and, in turn, packing the tins with ice, in shipping boxes.

We bought a large tin of each kind of meat and went back to the motel to visit for the evening. After quite a few convivial hours at this, anyone's constitution should have felt hammered, and mine played along with the odds.

The following morning, Ed woke up, rose up, showered and shaved and cheerfully said, "Let's go have some breakfast." I fell into an unenthusiastic lock step, even to the extent of the tall glass of acid grapefruit juice that removed the lining from my gullet and the greasy bacon and eggs and hash browns.

At this point, a gentle pink clergyman for whom Ed had once erected a church, stopped by and seemed to chat on interminably. Now I'm not saying that this preacher was sly, but I felt he was saying, "The wages of sin is death by hangover, Steve, and I'm going to string out this discussion so that I can watch you go."

There was so much brass and desperation mixed together in Ed that he was a hard guy to fathom.

He was a promoter who hatched some great ideas but once these ideas looked like they might go into successful production, his thoughts began to wander to new schemes. I do believe that, for him, the squeal was more important than the pork chops.

Later on, when I had a small struggling reman plant, he couldn't understand why I wouldn't spend money that I didn't have to paint the buildings and put the employees into uniform coveralls with the company logo on the back.

Ed and his erection crew built a golf club complex at Titusville, Florida, that is a visual delight for any lumberman. He also did many spectacular glulam beam and deck buildings when that look was in vogue, and I would rather remember him for these.

He has dropped from sight again, but he may reappear one of these days to give me hell and to add another story.

— * * * —

More difficult to read was the experience of a group of salesmen from one company who went out for a night on the town, only to have it turn to a near disaster.

As they hopped from bar to bar making new friends here and there, one of their members disappeared. They wound their way back to their hotel and, assuming that their friend was old enough to look after himself, took their plane out the following morning and went home without him.

After an alarming period of silence, word of his whereabouts finally came through. He had had a heart attack and was near death in a hospital in the Southern U.S. city they had been visiting. Fortunately, he survived, but his fickle friends had something serious to think about for a while.

— * * * —

Lumbermen's wives suffer in the service.

Something that is common to many men of my ilk is the irresistible urge to see what the other fellow is doing in his operation, and, of course, to talk lumber. I have taken this to the extreme more than once at the expense of my wife's patience.

On one occasion when we were visiting my sister in Porterville, California, we took a drive due east toward the Sierras just to see what was there. What was there, in this most unlikely place, was a sawmill. Of course I stopped to have a look, and, as so often happens in this business, the manager and I fell into an easy conversation about people we both knew in Donald, B.C. and in Spokane, Washington. Meanwhile, Gerry had to cool her heels, and I lost some more points.

Another time, we drove from San Jose to see Sequoia National Park. This took us up through the farms and vineyards of the San Joaquin Valley. At Dinuba, near Fresno, I found myself driving past the Ivory Pine Lumber Company, owned by Pat Ivory. Pat was the idea man of his day, having his "inventions" written up regularly in the Timberman magazine.

As far as the eye could see in all directions there was not a single tree. The company had an arrangement to do selective logging in the National Park and were also logging a large area of the park that had been killed by a forest fire.

My car refused to go past the gate, so I parked and went in to visit with Pat himself. Gerry remained in the shadeless car in 110 degree blazing sun while I spent a couple of hours looking at Pat's marvellous gadgetry.

When it was later pointed out to me, I had to agree that this was not very fair.

Chapter 10

Getting Down to Business

The biggest goof of my life turned out to be a blessing in disguise because Kicking Horse sent me on my way to the Coast and to Allwood Industries, a company that we built from an initial $500 investment in a hand-pulled trim saw into a twenty million dollar a year business that lasted twenty-four years (so far).

Allwood became the repository and testing ground for everything I had learned. To the experiences from that twenty-six year Odyssey, I added a mountain of new, interesting and educational ones in which I was ably assisted by excellent people, early and late, both in the plant and on the salaried staff, and by some excellent technical advisors who were the next thing to company employees. For eleven years I had a partner who will appear in his turn.

The magnificent accomplishments of the forest industry's heavy hitters are well-known to most people in B.C. Whether from having a plant in their city or neighbourhood, or from just reading the newspapers, most people recognize the names of the companies and the names of some of the people who run them.

It is no surprise to anyone these days that most of

the industry's enormous timber consumption flows through the facilities of eight or ten major companies. It was not always that way and I don't suggest that the business could or should have developed differently given, first, the anxieties over the chances of a perpetual timber harvest and, second, the acknowledged difficulties in approaching that problem from any other direction than that of an integrated company — one that produces pulp, paper, packaging, fibre boards, and plywood, as well as the lumber that still sentimentally lends its name to the business.

Obviously, when there was an abundance of easily accessible timber, many small companies took advantage of it, but one by one they disappeared. This suggests that a lot of entrepreneurs should also have disappeared with these resources, but fate didn't get them all. There is still a sub-culture, if you will, in the industry that is just as vital as "big" industry. It is just a matter of scale. Its members have to think twice as hard and work twice as hard in order to stay alive, because they control no raw material. They live at the whim of the major companies who can say at any time, "We're not going to sell you that item any more. We've decided to use it ourselves."

And they do, but for a major integrated company using its own raw material, remanufacturing lacks that certain quality of life or death urgency.

Because the people in small business are the industry's free spirits, they come up with a disproportionate number of bright ideas. If they feel like trying something, they just go ahead and try it. Approval for expenditure remains in the hands of the guy with the bright idea, so the chances of the project being done right are substantially increased.

But they are playing the nickel machines.

— * * * —

The coming of Allwood was an unconscious event. There was absolutely no planning involved. When I phoned a friend to say that I was coming back to the Coast, he said, "You have always talked about getting into the remanufacturing end of the business. Why don't you do it now?"

While I was at BCM, I had made a couple of furtive attempts to get into something, but they didn't go anywhere. I had a good job and the impetus to do something this daring was missing.

One of these earlier propositions is worth mentioning. I saw an ad in the paper for the sale of a reman plant, and looked into it. What the fellow had to sell was a resaw, a cutoff saw and a nailing machine that he had set up in a barn on Matt Kennedy's farm at Mud Bay. He was making lettuce crates for Kennedy and lath and dunnage for the waterfront, but he was doing pathetically. I remember that he gave me a little balance sheet and operating statement, and listed in his assets was a value that he had put on the orders he had on the books. These were going to make money once they were produced. I suppose that was goodwill, but there wasn't a stick of raw material in the yard and he was deep in debt to the bank.

The fellow's banker was in trouble with his own people for letting the fellow get in so far, but he was honest with me. I said, "What would you think about me taking this thing over?" He said, "I think it would be wonderful, but I can't imagine why you would want to."

There has been a lot of money made since, doing exactly what that fellow was doing, but you have to stay sober and you have to be persistent.

— * * * —

After you have knocked around for a few years and met a lot of people and tried a number of different things, you are either a find for someone who needs help, or you might just be so opinionated that you are unemployable.

I thought that this latter might be the position I now found myself in. I knew I could do the things I had been doing and that I could adapt this ability to just about anything that came along, but I also knew I was getting skeptical about some of the things I saw going on around me, and that, to satisfy myself, I was going to have to break new ground.

That may sound like wisdom after the fact, and I don't suppose I ever did say, "Tomorrow morning I must start my own business," but circumstances gave me a push and this time I didn't chicken out.

What followed was a lesson that only time can teach. No matter how much you know in general, you never really know it until you do it yourself. When you do learn, you look back and say, "My God, I was dumb. Why didn't I do this instead of that?" And the answer is, "Because I didn't know any better at the time." No matter what you learn and do elsewhere, your own operation is always going to be different in some respects and this will be all new ground.

I never worked so hard in all my life, even on a Fraser Mills green chain when the lumber was two feet deep and my piece was on the bottom. The jump into actually doing business, buying and selling and manufacturing and having to answer every question yourself, is not a thing you are good at right away — but in my opinion it beats any other kind of work there is.

Hindsight is not much help unless you can live two lifetimes, but you do look back on some things and groan.

The worst handicap that you can saddle yourself with is starting too small and struggling to make money with a production level and cash flow that is inadequate to support the bare overhead necessities and a salary to live on. The energy spent robbing Peter to pay Paul is exhausting and non-productive. If the little business staggers, the blame is often misplaced. You question your methods and your margins and even the credibility of the business idea itself, when in truth all that is missing is the capital to get everything in the right proportions at the start, and to have a cushion for the inevitable stuttering startup. I said "capital" instead of financing on purpose. Interest paid is someone else's profit and if there is any profit to be made, it should be yours.

How do you know that this is the problem? After a year, or five years, or even ten years of struggling and growing, you finally get there and you find that with the right equipment and the right staff and enough money to operate, you are making a profit doing essentially the same things you were starving on when you started out. It looks so dumb on the surface that you would think starting on a shoestring would be illegal. Unfortunately, most of the time there isn't any other way, because all bankers and most investors only want to back sure things. If you want to gamble, you have to do it yourself.

The thing that perpetuates the shoestring startup is the determined misfit with more energy than either money or experience, or perhaps even brains, who is prepared to blunder his way to success. The woods are full of them.

Did you say that Professional Engineers and MBA's and Ph.D's would be too clever to suffer through that torture? Well, degrees don't guarantee anything. Money and markets and plant and connections and trained and weeded staff and time and timing and energy and luck and money and brains — these are what guarantee success.

Having brains enough to get work done effectively

by others is quite sufficient, and brains of that kind don't come in a package with an identifying label.

Experience, guts, and judgement, tempered by an agreeable personality — now that has a nice ring to it. Add to that a kind but absolute authority, and you are getting close to a successful formula for a small business operator. SOMEBODY HAS TO BE BOSS AT ALL TIMES.

When I decided to groom someone to succeed me at Allwood, I picked a very strong person. When that person built up experience and self-confidence and convictions to the point where the company no longer had a clear-cut leader, it was necessary, but fortunately easy, to put the company back into the hands of a single decision-maker. At my tender age it wasn't going to be me. The new boss will send the company on its way to a new life, with a new image, like a piece of reconditioned machinery. Time will tell if the new direction will be better than mine.

But that was 24 years later.

— * * * —

All this talk about autocratic rule is going to be questioned because a lot of people know that I had an active partner with me on the site from 1971 to 1979 and that we remained financial partners until 1983. I should clear this up right now. Jim Fenton and I divided the company into sales and production functions and we respected each other's areas, but I always deferred to the necessity of a full order file and adjusted my prejudices accordingly. Policy was arrived at through consensus but since we were equal shareholders, we had every good reason to cooperate. Besides this, we knew a number of things and many people in common, and had travelled over different but similar routes when we joined forces, so our ideas made a pretty good fit.

Even though the latter half of the seventies saw good times for lumber manufacturing, we still ran a Spartan organization, and every key person had at least two jobs. We did not load up the company with assistants with fancy titles, nor did we scramble for a spot on the front page of the newspaper. Our customers knew who we were and that was all that mattered to us.

Jim's philosophy boiled down to this: "Never turn your back on a problem." He lived it fiercely and faithfully and he was better at it than I ever was. But it made for long days.

Along the way we built Central Cedar, a substantial lumber distribution yard, and got it on its way before we broke the company into two separate entities and went our separate ways.

— * * * —

To get back to our shoestring start, the property we bought in 1965 was the triangular tip of a major industrial development. It still had a few trees, a spring, a lot of skunk cabbages, and a gravel road along its frontage. And we got it cheap. Today people ask, "How did they ever let that 'sawmill' build right in the centre of town?" The answer is that it was a small town and we were on the outskirts at the time.

We started out on this piece of bog by setting up our little trim saw at the edge of the field, so to speak. The trimmerman had a tin roof shelter over him but everyone else worked in the rain. My wife Gerry manned the office, I hired two strangers off the street, and we started to trim Texas fencing. The first car was sold to Phil

Henderson at Hugh Brady Lumber in Seattle, and they sold it to Slaughter Bros. in Dallas. The second car went directly to Slaughter, and Al Slaughter and I kicked Allwood off.

Within the year we had built the first section of a factory building out of treated poles and trusses, and we got down to the serious business of growing up.

— * * * —

There was a bulldozing contractor in Langley named Jim Hassell. He cleared our land and did the first gravel filling. Jim had lost his larynx to cancer and he burped his words. One day he put his arm around my shoulder and rasped, "Steve, let me give you some good advice. Never keep your own books. You'll lie to yourself."

As mad as I have been at C.A.'s who seem infuriatingly determined to present the worst possible picture, I have never had skeletons in closets that could pop out and shatter a creditor's confidence.

— * * * —

In its first 24 years, Allwood saw good times and bad times. It was off to a struggling start in 1965; began to grow by overextending (i.e. using precious working capital to build plant and thereby perpetuating the cash drought); it was staggered by the recession in 1971, thrived between 1972 and 1980. It was burned to the ground in 1981, and rebuilt in time for the rest of the 1980's recession, when it nearly failed. It recovered again, but a grim experience such as we went through teaches you to say "Gee, let's not do that again." We saw better times from 1986 on, but the memory of the bad times permanently changed our attitude to a number of things.

For me, Allwood Industries was the culmination of many years of looking for the elusive ways to make an unusable piece of lumber useful, and in the process to add more dollar value to the resulting products than was expended in added production costs.

I'd had some experience with this process. The first grading job I had at Fraser Mills was in their remanufacturing and recovery department.

At Kootenay Forest Products we bought pine lumber from outside suppliers and made money processing it. As a matter of fact, sawing eight species with each one having its own distinct products was the ultimate preparation for remanufacturing.

At Powell River/MB we looked at shingle logs from every angle to try to recover lumber profitably. We also used the remnants of the old box factory as a reman facility to make furniture parts, flush door cores, gothic-pointed fence pickets, and a host of other items.

Every major sawmill in the industry did a total manufacturing exercise from log to finished product up until about 1960 when the traditional finished products would no longer return their production costs. At this time, the major mills began to concentrate more on industrial grades to sell for processing elsewhere, where costs were cheaper. They identified the point where touching the piece again would reduce its contribution to the overall profit from the log and they stopped there.

There is no doubt that the remanufacture of lumber is fascinating, challenging and satisfying to people who have the right background and are not afraid to get a little sawdust on them. It is a "hands on" business, if I may use that hackneyed term. It is also a skill-testing exercise to find yourself wedged in between your suppliers and an inflexible union contract without enough weight to move either one of them. The fact that B.C. costs ultimately got so skewed that Allwood's "low value" corner of the business was driven underground into hole-in-the-wall non-union plants, or out of the country to U.S. manufacturers, was one of the major frustrations of my life.

Finger-jointing, edge-gluing, cut-to-size components, and the full range of furniture and mill work products have been around in B.C. for a long time. Selling the finished products from these manufacturing operations to countries that can do the same things cheaper has become the magic trick for B.C.'s secondary industry to master.

Unit cost is a function of volume and where the product to be recovered is small and unsuited to automated procedures, with a resulting high labour cost per piece, the product often cannot compete with the alternatives, and the proposition founders. Under these circumstances, some of the intriguing specialities that could be made can't be made. This results in an anomalous situation where remanners are forced to return to simple, straight gut production programs in order to stay in business. Unless we change these circumstances, we give the lie to the romantic notion that secondary industry is the future of our country. In B.C., even the essential elements of a remanner's variable costs — his whole operating agenda — are overshadowed by an inflexible union edict that says "What's good for MB is good for everybody," i.e. an industry-wide master agreement that puts everyone, big and small, in the same boat.

By this rigid system, people with good ideas are often prevented from doing good things for themselves, for new employees, or for the country, because they are trying to compete on a tilted playing field.

— * * * —

Having got that off my chest, I will return to my original contention that people and the things they said and did made my life and my work enjoyable, and that sustenance came along for the ride. Sure it seemed frustrating and barely bearable at times, but at other times it was an unharnessed gallop through life that left me a young man in a thoroughly used body, wondering how it all could have happened so quickly.

— * * * —

Like all the other places where I had worked, Allwood attracted its share of characters.

Since we were such a small business at the start, I was very close to the employees. In those early days, I was the extra man in every crew and did every job in the plant. After we were unionized one of the men actually went to the union and said, "How come Steve works? He's not in the union." Strangely, the answer was, "He owns it so don't push that question." They would not always be that broadminded.

— * * * —

Allwood walked the straight and narrow, so we neither got very rich nor thrown in jail.

There was an episode once when I custom-planed a series of truckloads of rough lumber and it was developing into nice steady business. Then one day it stopped abruptly.

A serious-looking gentleman with short hair and a clipped moustache walked in and asked me what connection I had with the lumber I had been planing. When I told him, he gave me his card. He was with the RCMP Fraud Squad. The lumber was being systematically stolen from a mill and brought to our place for planing. The ringleader in the deal was well-known and actually did flee to South America. As far as I know, he is still there.

— * * * —

If you hung around Allwood long enough, you got to do just about everything.

Somewhere in the middle 1960's, the farmers in the Fraser Valley switched over completely from pole beans to a low bush bean that could be harvested mechanically. My next-door neighbour was a foreman for a major canning and freezing company, and he tipped me off on the bean situation. There were 7-foot hand-split cedar bean poles stored in piles all over the valley, none of which would ever be used again.

I went out and bought every one that I could get my hands on. We trimmed off the broken ends and the ends that had been in the ground, and sold several carloads of them as grape stake fencing. I recall that one car went to Virginia and another into Denver, Colorado. When they were gone, they were gone, but for a month or two afterward I got calls asking if I had more.

— * * * —

For a few years we had a pair of owls that lived up in the trusses in an abandoned section of blowpipe. They raised their young there and we thought of them as good housekeepers — they kept the pigeons away. It always gave me a tingle to come in early in the morning and see an owl leave on those perfectly soundless wings.

— * * * —

Over the years there were several car enthusiasts who spent every dollar they could spare on their wheels. One kid bought a 20-year-old beauty that sparkled like a diamond. One day at the end of the shift, as these men-children were jostling their way out of the plant, one of them, on walking past this car, said, "Yes, Dale, it sure is a beaut." With that he gave the fender a vigorous thump with his hand, causing a large patch of Bondo to fall to the road, leaving an ugly scar in its place.

— * * * —

An employee who has over twenty years in with the company was harum scarum when he arrived, and has shown no sign of changing. His track record for buying cars was dismal, but one day he arrived with an older blue Mustang that, for once, looked to be in very nice shape. He was so tickled with his purchase that he asked the superintendent to get in for a test run.

They headed down Logan Avenue with a roar and in a cloud of smoke from burning rubber, but when they had gone a couple of hundred yards, stuff started falling off underneath the car. The drive shaft and other assorted parts hit the road, and sparks were flying everywhere.

Once the driver had overcome his shock, he brought the car to a halt in the middle of the road. It sat there very still and quiet for a moment, then the two of them got out and pushed it off to the side and walked back to the mill, to be greeted by a rousing chorus of catcalls and raspberries.

— * * * —

There is an old story about the Spartan soldier who tamed a fox and kept it as a pet, against the camp rules. One day he was surprised by an officer while he was holding the fox and to excape detection, he hid the fox in his cloak. While the unflinching soldier stood at rigid attention, the frantic animal gnawed at his vitals. Finally, the soldier fell dead.

We had a Spartan at Allwood Industries. He was the above mentioned collector of used cars. One year, on the day after Halloween, this fellow thought he would have some fun at Jim Peterson's expense. He drove his forklift close to where Jim was working then produced and lit a small firecracker. This he planned to drop behind Jim just to see him jump. Just as he lit it, I walked around the corner and caught the prankster with the sputtering firecracker in his hand. He hurriedly attempted to snuff out the fuse but he was too late. At the last second, trying desperately to hide it, he closed his hand over the firecracker and let it explode.

— * * * —

We had a coffee truck that called on us in the early days. One day, as Lil, the driver, was approaching the plant, a universal joint came apart and one end of her drive shaft dropped to the road. She was just able to coast halfway into the plant driveway.

Logan Avenue was still a gravel road, and a fellow who had just bought a horse at the Thursday farmers' auction was riding down the side of the road at a full gallop.

He was riding bareback and the horse was obviously out of control. They headed straight for the protruding end of Lil's coffee truck, but at the last minute the horse saw what was up and veered to miss it. The rider's progress observed the law of tangents and he flew on straight ahead. He hit the gravel and slid right under the truck. The natural reaction was to get up and get out of there, but all that he accomplished was to hit his head under the truck and drop back to the ground.

It was a scuffed and sheepish looking cowboy who finally stood up with a classic "Where am I?" look on his face.

The whole episode lasted an exciting thirty seconds.

— * * * —

Once I bought an ancient Hyster forklift in Port Kells and drove it back to the plant. At that time, 200th Street was still gravel and the hill coming down into Langley had not yet been cut down. It was steep. I soon found that the machine had hardly any brakes. The king pins were so sloppy that I was weaving all over the road as the steering wheels flopped from one side to the other.

As soon as I got over the top of the hill I knew I was in trouble. Somehow, I got stopped and turned the machine around and backed it down the hill in reverse. The reverse gear range was low enough that the engine braked for me; otherwise it would have been a case of "jump and watch where it goes."

We tore up the steering box in this old bucket one day and the only thing we could find to fit was a steering box from an International truck. Now, trucks steer with the front wheels and forklifts with the back, so the worm gear in one steering box is opposite to the other. This posed an unusual problem. The forklift driver now had to turn the wheel to the right when he wanted to go left, and vice versa. We ran that old thing like that for two or three years. Stan Tinkess, the driver, said it was O.K. once you got used to it. We all expected to see Stan jump in his car one night and drive across the road and into the field. Worse than that, he drove race cars at the Langley Speedway. We eventually bought a brand new Clark and I sold the old timer to Rod McKerlich, who got another ten years out of it.

— * * * —

Not all the Allwood stories were happy. In about 1970 we hired a fellow who soon showed that he was overqualified for the jobs we assigned him. He was articulate and intelligent. Whimsey was his middle name and we just couldn't believe what finally happened to him.

One day, he was the inside man of a car loading operation and I jumped into the car and took every other piece so that I could talk to him. I asked him what his background was and he said he was a social worker.

"Some people can be social workers and some can't. I can't," he said. "I put too much of myself into the job and when my charges lied to me and reneged on promises and commitments, it ground me down."

As it soon turned out, he was not without his own problems. He lapsed into serious drinking and lost time because of it.

One day he said he was going to be gone for a couple of days as he was the only family member who would attend to the arrangements for his grandmother's funeral in Kamloops. He was missing for several days and they found him and his car at the bottom of a cliff in the Fraser Canyon.

When his father-in-law came in to pick up his final cheque, I asked him about the Kamloops funeral. He said, "Oh, he didn't have any grandmother in Kamloops."

— * * * —

No one could accuse Allwood of being equipment proud. When we added anything new, the first place we looked for the makings was in the bone yards of the major mills. We got chain, shafts, sprockets, rolls, whole sections of chain transfers, belt-conveyors, motors, gear reducers — you name it — all discarded and replaced by shiny new gadgets. We scrounged through this stuff, adapted it, and sent it on its way to another twenty years of honest service. I'm sure the lumber didn't care if it was travelling on new or used equipment. The main thing was that it knew where it was going and, at Allwood, there was never any doubt about that. We hatched some clever innovations but we didn't build any white elephants.

We never had room to build anything grandiose, but we do think we originated two concepts that are in common use today, though, having done a couple of patent searches in my day, it is possible that even these were thought of by someone back in 1920. One thing is sure: we got them out into the open.

One is a hydraulic tilt hoist that operates from floor level without a pit, and the other is the push-out roll setup on planer and green chains that replaced carriers and carts. We had this setup in the early 1970's, and as far as we knew, it was original.

One of the slickest things we ever built involved an air cylinder with a greasable ball joint at either end that allowed the cylinder to push in any direction. The ball joints were Ford farm equipment — a harrow hitch or something equally sophisticated.

We built a declining belt transfer that also travelled on an angle from right to left, and allowed us to establish a lumber line just where we wanted it.

We built a planer chain that was twenty feet wide at one end and eight feet at the other, just to steal a little space. It worked like a darn and ensured that the guy pulling the long lengths never went to sleep.

Even when I was at KFP we bought a derelict grease rack from a service station that was being demolished. It cost us a hundred bucks. With a little imaginative fixing, it made an excellent accumulator hoist. It sloped so it looked a bit like a depth charge thrower, but it worked perfectly.

Unsophisticated? You bet.

Effective? Right again.

— * * * —

Early one Sunday morning, April 26, 1981, to be exact, my home phone rang and the operator at the RCMP office advised me that our plant was on fire and that I should get out there as soon as possible. That is a call everyone can do without. By the time I got there, it was almost gone and it brought a big lump to my throat. Allwood represented sixteen years of growing in the classic lumber plant tradition of adding on and leaning to, but there were some wonderfully simple ideas there that couldn't have worked better for our purposes.

We thought briefly about turning it into a used car lot, but once we decided to go back in business, the activity was furious. We had a building completed and a planer operating by October and by the end of the year, the whole place was going again with the best ideas of the past all incorporated into a unified plan. We did our planning so well that in the ensuing seven and a half years, we only managed to make two or three additions and revisions, a rather slow pace for us.

The fact that Canada, the U.S., and probably the world were treating themselves to a major recession took some of the fun out of breaking in the new plant. Before the recession was over, we had asked ourselves a hundred times why we ever bothered to rebuild.

When things got tough the bank retreated to its cold world of probabilities and wrote us off.

The union had an agenda of its own that didn't include us, but in their case, I'm sure they had their fingers crossed under the table.

In the end, it was our suppliers in the industry, the only ones who really knew what was going on, who trusted us and stuck with us.

We learned a lot from this nightmarish episode and my young proteges saw at first hand the anxiety of nearly going broke.

— * * * —

A reman plant is not the Brookings Institute. We had a repeat of the "that's my place" situation at Allwood. We had a forklift driver named Mark who was a gentle giant but a very powerful one. Anyone who would even consider aggravating him would be out of his mind.

Mark had parked his car in the same spot in front of the lunchroom for years, but one morning he came in and found a well-known car in his space. Mark went into the lunchroom and asked the usurper to move his car. The other fellow was feeling brave and made a wisecrack. Mark lunged for him. The guy threw his coffee at Mark and then ran out the door, into his car, drove right out of the plant, and didn't come back that day. This left the rest of the crew, including Mark, splitting their sides with laughter.

— * * * —

Once a kid named Harder was an extra man for a day so I asked him if he was a carpenter. He said, "Well, sort of, I guess," which didn't tell me much. We had just enlarged the lunchroom and I asked him if he thought he could build a table. His answer was such a confident "Yes" that I turned him loose without another thought.

As it turned out, Harder was not a carpenter, but he had seen a picnic table in a park once. When I called around to see how he was doing, a picnic table was what was underway.

It was big and it was ugly, which didn't surprise me, but what was alarming was that Harder didn't seem too familiar with the dimensions of the average human being. When he got it finished and the crew came in to give it a tryout at lunch, they found that the top was so high that short men could barely see into their lunch buckets.

When this was brought to Harder's attention, he assured everyone that that would be easy to fix. He rebuilt the seats so that everyone could see over their lunch boxes and across the table, but now nobody's feet could reach the floor. It lasted that way for a couple of years until we added on to the lunchroom again and built a fine long table with an arborite top and individual chairs.

— * * * —

We had a fellow named Joe who was a genuine eccentric. As out of touch as he seemed to be at times, he still had accumulated some private wealth. He had a few houses and a few tenants for whom the only rules were Joe's rules. He had a disagreement with one tenant, and to bring the matter to a head, he loaded a rifle and went to the house where he shot a hole in the tenant's washing machine.

Joe had to work in close proximity with other men, but most of the time he came to work reeking with such a disagreeable smell that no one wanted to work near him. The answer was not the usual personal carelessness you would expect. Joe kept goats and the goats liked Joe. The billys claimed him as their own by marking him with their scent glands and it was his clothes that reeked.

I once saw a silent movie comedy going on out at the green chain. There was no lumber coming from the saw and Joe was taking the opportunity to clean up the odds and ends that accumulate at the end of the chain. He stopped the chain to do this work but just when he would start to look organized, someone up the chain would start it again and bury Joe's good works in a deluge of new lumber. The next thing I saw was Joe holding his tormentor by the lapels of his jacket and shaking him until his eyeballs rattled. In view of Joe's tenuous grip on reality I feared the worst, so I had to run out and negotiate yet another peace treaty.

One day Joe came into the lunchroom and removed a gold bar from a neat little cloth bag. I don't remember the weight, but it was the size and shape of a jumbo-size chocolate bar, and on that day's gold prices it was worth $5,000. He passed it around and everyone agreed that it felt "real good".

— * * * —

We had a night shift foreman named Ben Zilkowski. One night, as I was sitting down to supper at home, Ben called and in a woeful voice said, "Gee, Steve, I've done something terrible. I drove into the post holding up the main beam in the dry shed and broke it right off." I was prompted to ask, "Golly, Ben, what's holding the roof up?" "Gosh, I don't know," was his artless reply. "Well you'd better get out there and have a look at the effing thing, if there is anything left to look at."

The real answer to my question was: "Nothing but some plywood scabs over the joints in the purlines." A 12"x 30" glulam beam was hanging from them — all of this with no visible means of support. We rescued it by supporting the beam with the top of the raised forklift mast until morning, when good hands did a major rescue job.

That was not the only time we had a close call. A driver once dislodged a huge beam over a main doorway and all that held the trusses up was the strapping and roofing. When we rebuilt after the 1981 fire, we placed all the beams so high that no forklift could reach them — unless the driver could make his machine jump.

— * * * —

Like the malapropist who finds a "flaw in the ointment," the antics of some characters are almost too pat. You wonder if they are hams, playing to an audience.

Jim Peterson was a natural. He was a scrawny little man who clumped around in steel shod boots that seemed to weigh as much as he did. He always looked and sounded serious, but his stories were pure "Munchausen."

He thought nothing of quoting more than one birth date. It depended on what was needed to qualify him for the benefit of the moment. When pressed for a birth certificate, he brought a letter from his wife stating that she knew how old he was. He was born, etc."

He got a leave of absence to go to the funeral of his stepson in Ontario. The following year he made the same request. When I protested that this same fellow had died the year before, he said, "Oh, we thought he was going to die last year, but he didn't. This time he died."

He came to me before the shift one morning and asked for the day off to go to a funeral — in Revelstoke, a nine-hour drive.

He went to his doctor and got a cortisone shot for a sore elbow. The doctor told him he was putting oil in it so it wouldn't be so stiff. Jim repeated this to the lunchroom crowd in all seriousness.

We had a plant cleanup man who boarded at Jim's house. This fellow was an awful old reprobate — the kind who doesn't wear his teeth.

One Saturday morning Jim Fenton and I were in the plant and we heard someone rustling around the corner of the building out of sight. It was Peterson. When we asked him what he was doing, he said, "That damned Art came home drunk last night and I got so mad, I threw him right through the wall. I'm taking this little bit of shiplap to patch up the hole."

Jim's wife's name was Ruby. Ruby would phone and ask my wife Gerry to go out into the plant right away to fetch Jim — she had an emergency. Jim would rush home and return an hour later to explain that the pigs had got out.

One incident that was true and that Jim Peterson developed into a full-fledged saga concerned a time when another fellow was going to burn a pile of scrap wood at the end of the yard. He threw gasoline on the pile and lit it, with the predictable result. The gasoline fumes exploded and the fire builder suddenly found himself enveloped in flames. It had been raining and Jim's version was that he grabbed the victim and rolled him in a puddle until the fire in his clothes was out. Jim weighed about 130 pounds. The other fellow was closer to 180, and I'm sure he was only too willing to roll in the water, but we accepted Jim as a genuine hero anyway.

— * * * —

One day we got caught in one of those heavy wet snows that comes suddenly and fouls up everyone's trip home from work.

Jim Peterson, anticipating a problem, had someone set a unit of lumber into the back of his pickup to give him some weight for extra traction. The following day he came back with the truck empty and had to be reminded of something he was hoping we would forget.

Another character working a night shift was about to leave with his pickup full of bales of insulation brought in for a lunchroom addition. His defence was, "I didn't steal anything. My truck and the stuff were still on company property!'

Another guy who wasn't much smarter loaded his pickup with lumber right off a planer run. It was an oddball beaded siding which was being run for the West Indies and it was all identified with grade stamps and lot marks.

On the way up the 200 Street hill, both back tires of the pickup blew out and left it looking like a ship sinking by the stern at 3:00 AM. The thief was duck soup for a passing RCMP cruiser. His story to the policeman was not recorded, but it was probably a dandy.

— * * * —

Looking back on what I have written here, one might get the impression that I just skimmed over the surface of the lumber industry for 50 years. Nothing could be further from the truth. Anywhere I was employed, I started each day early and stayed until the day's problems were as solved as possible. I always felt personally responsible when financial results were disappointing and this ground me down a bit. Anyone who doesn't feel that way probably feels that he isn't required to, since he is only working for wages, but he would be a better person for knowing the score and taking more interest. If he and others with him did, their companies would do better and it wouldn't take them long to realize that working for a healthy company beats working for a struggling one any day.

This Utopian concept may not happen until we move up to the next plane where adversarial labour relations are replaced by a form of profit sharing, and where the proposition that the smarter you work, the better you do, would make unions redundant. My personal philosophy could have handled this arrangement in my small company, but the union got to me too early in the game and wouldn't even let the men try it when it was offered to them.

Sometimes unions can be the most rednecked reactionaries of us all. They are terrified of change and are still back in 1939, saying, with Maurice Ryan, "If I give it to you, I'll have to give it to everybody." They cannot conceive of different plants and different products having different economic chances, nor will they believe that, sometimes, half a loaf is all there is and the choice is between that and shutting down. Competitiveness has nothing to do with it. From their idealistic viewpoint, one size fits all.

— * * * —

At Allwood, I was totally involved in the daily job of buying material, manufacturing, and coordinating those activities with the orders the sales people could get or with the specifications their customers were asking for. That, along with the administration and financial concerns, was a full time job and typical of a small business operator's diet.

There were peoples' whims and personalities to accommodate — the continued puzzle of how far was too far when it came to stretching the rules, and how to succeed in the fine art of balancing the need for control on one hand with contentment in the ranks on the other.

After writing the above, I came across the following in a book about the World War II Canadian Navy:

"We did it the hard way in small ships, without the help of big ship routines and personnel; we learned to strike a balance between the discipline necessary to run an efficient ship and the easy tolerance essential to the close quarters of our tiny vessel."

I couldn't have said it better, but then I once lived that life at first hand.

It is true that, if I had been tougher, I would have missed or cut short the careers of a few characters who raised absenteeism and its convoluted explanations to a fine art. I would have missed some wonderfully creative requests for "drags" to tide people over until the next payday. These were "the-dog-ate-my-wallet" type of things, and there should have been a monthly prize for the best story.

There were garnishees, suspended driver's licenses (one fellow came to work in a taxi every day for what seemed like years), broken marriages, broken hearts and broken noses. When you are a small company, you see it all up close. But as counterproductive as some of these things were, in a perverse way they gave the company a heart and soul and a sense of humour.

My wife always knew when I'd had a trying day and her reaction was invariably, "Why do you put up with those idiots?" This immediately sent me over to the other side to say something defensive that implied that they might be idiots but they were my idiots.

I have a friend who was a small contractor, building

beautiful custom homes one at a time. He echoed the above sentiment. "Sometimes they drove me nuts, but on the whole they worked their hearts out because they wanted us to be successful and to keep doing what we were doing."

The use of the word "idiot" is fraught with danger. To get the context established and perhaps even point a finger, I will tell one more story. For a while it seemed that booze may have left the scene to be replaced by the strange exotic substances of the modern world. A few of the young fellows got into pot early on, and one of them came to work one morning as high as a kite on something. It was not the first time, so I hustled him out and told him not to come back. He left, but returned a few minutes later to ask if he should put in for his 2 hours' call time pay.

But, most of them were good guys whom I also considered to be good friends. Several had been with me for over 20 years when I sold the company.

— * * * —

Epilogue

I suppose we could all look back and second guess some of the moves we've made in our lives.

If I could back up the whole sequence of events to a particular happy point and stop there, the things that occurred later on would have to be eliminated from my chronicle. And, since I wouldn't have missed some of them for the world, I would rather say that the right events probably happened in just about the right order. My aim here was to put these once-only experiences into perspective and let the events speak for themselves.

The activities of the major companies were a world removed from mine, but I dealt with all of them and gave them substantial amounts of business. It was a major source of satisfaction for me to be recognized as an important customer after the modest start we had made. And it is an amusing thought that tomorrow some new small operator will approach the sales department of a major company to ask if he can please be a customer. Let me tell you something: I wish it was me.

— * * * —

xxxvi) The original Allwood Industries plant, Langley, B.C., before the fire.

xxxvii) Original Allwood Industries plant. One add-on at a time. The mice got at the lumber in the foreground.

xxxviii) Allwood Industries after the fire, 1981.

xxxix) New Allwood Industries plant, Langley, B.C., after 1982.

xl) Interior view of new Allwood Industries plant. Four moulders, seven recovery trim saws, band resaw, four sorting tables, etc. A good plant.

xli) The new guard at Allwood Industries.
L to R: Kathy Zucchi, Colleen Gellein, Michael Holob, David Sasaki

About the Author

Steve Gatensbury was born in 1923 in New Westminster, while his parents were living in the Canadian Western Lumber Co. townsite at Fraser Mills, British Columbia. He started to work in the sawmill industry in 1939 at 16 years of age, and sold a successful remanufacturing business to retire in 1989.

He says he must have arrived in the industry just after the good old days, and left as computers were replacing the hand skills, the sweat, and the experience based, puzzled-out decisions that moulded yesterday's lumbermen.

In this book he tells of some of the things that happened in the intervening 50 years.

Steve lives with Gerry, his wife of 43 years, in Tsawwassen, B.C.

— * * * —

PHOTO CREDITS

The photographs reproduced herein numbered viii, x, xi, xii, xiii, xiv, xv, xvi,xvii, and xx are courtesy of the Vancouver Public Library.

All other photos are from friends and from the author's private collection.